HUNTING
EL CHAPO

ALSO BY DOUGLAS CENTURY

BARNEY ROSS

TAKEDOWN

STREET KINGDOM

ICE

IF NOT NOW, WHEN?

BROTHERHOOD OF WARRIORS

HUNTING
EL CHAPO

The Inside Story of the
American Lawman Who Captured
the World's Most-Wanted Drug Lord

Andrew Hogan and Douglas Century

HARPER

An Imprint of HarperCollins*Publishers*

HarperCollins books may be purchased for educational, business, or sales promotional use. For information, please email the Special Markets Department at SPsales@harpercollins.com.

FIRST EDITION

DESIGNED BY WILLIAM RUOTO

Library of Congress Cataloging-in-Publication Data has been applied for.

ISBN 978-0-06-266308-5

18 19 20 21 22 LSC 10 9 8 7 6 5 4 3 2 1

AUTHORS' NOTE

This is a work of nonfiction: all the events depicted are true and the characters are real. The names of US law enforcement and prosecutors—as well as members of the Mexican military—have been altered, unless already in the public domain. For security reasons, several locations, makes of vehicles, surnames, and aliases have also changed. All dialogue has been rendered to the best of Andrew Hogan's recollection.

To my wife and sons.
—A.H.

Certainly there is no hunting like the hunting of man and those who have hunted armed men long enough and liked it, never really care for anything else thereafter.

—Ernest Hemingway, "On the Blue Water," 1936

CONTENTS

CONTENTS

PROLOGUE: EL NIÑO DE LA TUNA

PHOENIX, ARIZONA
May 30, 2009

I FIRST HEARD THE legend of Chapo Guzmán just after midnight inside Mariscos Navolato, a dimly lit Mexican joint on North 67th Avenue in the Maryvale section of West Phoenix. My partner in the DEA Narcotic Task Force, Diego Contreras, was shouting a translation of a song into my ear:

> *Cuando nació preguntó la partera*
> *Le dijo como le van a poner?*
> *Por apellido él será Guzmán Loera*
> *Y se llamará Joaquín*

"When he was born, the midwife asked, 'What are they gonna name the kid?'" Diego yelled, his breath hot and sharp with the shot of Don Julio he'd just downed. "The last name's Guzmán Loera, and they're gonna call him Joaquín . . ."

1

Diego and I had been working as partners in the Phoenix Task Force since early 2007, and two years later we were like brothers. I was the only white guy inside Mariscos Navolato that May night, and I could feel every set of eyes looking me up and down, but sitting shoulder to shoulder with Diego, I felt at ease.

Diego had introduced me to Mexican culture in Phoenix as soon as we met. We'd eat birria out of plastic bowls in the cozy kitchen of some señora's home that doubled as a makeshift restaurant and order mango raspados from a vendor pushing a cart across the street, all while listening to every narcocorrido* Diego had in his CD collection. Though I clearly wasn't from Mexico, Diego nevertheless told me I was slowly morphing into a *güero*—a light-skinned, blond-haired, blue-eyed Mexican—and soon no one would take me for a gringo.

The norteño was blaring—Los Jaguares de Culiacán, a four-piece band on tour in the Southwest, straight from the violent capital of the state of Sinaloa. The polka-like *oompa-loompa* of the tuba and accordion held a strange and contagious allure. I had a passing knowledge of Spanish, but Diego was teaching me a whole new language: the slang of the barrios, of the narcos, of "war zones" like Ciudad Juárez, Tijuana, and Culiacán. What made these narcocorridos so badass, Diego explained, wasn't the rollicking tuba, accordion, and guitar—it was the passionate storytelling and ruthless gunman attitude embodied in the lyrics.

A dark-haired waitress in skintight white jeans and heels brought us a bucket filled with cold bottles of La Cerveza del Pacif-

---

* Narcocorrido: a ballad in a traditional Mexican musical style whose lyrics recount the exploits of drug traffickers.

ico. I grabbed one out of the ice and peeled back the damp corner of the canary-yellow label. *Pacifico*: the pride of Mazatlán. I laughed to myself: we were in the heart of West Phoenix, but it felt as if we'd somehow slipped over the border and eight hundred miles south to Sinaloa. The bar was swarming with traffickers— Diego and I estimated that three-quarters of the crowd was mixed up somehow in the cocaine-weed-and-meth trade.

The middle-aged traffickers were easy to spot in their cowboy hats and alligator boots—some also worked as legit cattle ranchers. Then there were narco juniors—the new generation—who looked like typical Arizona college kids in Lacoste shirts and designer jeans, though most were flashing watches no typical twenty-year-old could afford.

Around the fringes of the dance floor, I spotted a few men who looked as if they'd taken a life, cartel enforcers with steel in their eyes. And scattered throughout the bar were dozens of honest, hardworking citizens—house painters, secretaries, landscapers, chefs, nurses—who simply loved the sound of these live drug balladeers from Sinaloa.

Diego and I had spent the entire day on a mind-numbing surveillance, and after ten hours without food, I quickly gulped down that first Pacifico, letting out a long exhale as I felt the beer hit the pit of my stomach.

"Mis hijos son mi alegría también mi tristeza," Diego shouted, nearly busting my eardrum. "My sons are my joy—also my sadness. *"Edgar, te voy a extrañar,"* Diego sang, in unison with the Jaguares' bandleader. "Edgar, I'm gonna miss you."

I glanced at Diego, looking for an explanation.

"Edgar, one of Chapo's kids, was gunned down in a parking lot in Culiacán," Diego said. "He was the favorite son, the heir apparent. When Edgar was murdered, Chapo went ballistic. That *pinche cabrón* fucked up a lot of people . . ."

It was astonishing how Diego owned the room. Not with his size—he was no more than five foot five—but with his confidence and charm. I noticed one of the dancers flirting with him, even while she was whirling around with her cowboy-boot-wearing partner. Diego wasn't a typical T-shirt-and-baggy-jeans narcotics cop—he'd often dress in a pressed collared shirt whether he was at home or working the streets.

Diego commanded respect immediately whenever he spoke— especially in Spanish. He was born on the outskirts of Mexico City, came to Tucson with his family when he was a kid, and then moved to Phoenix and became a patrolman with the Mesa Police Department in 2001. Like me, he earned a reputation for being an aggressive street cop. Diego was so skilled at conducting drug investigations that he'd been promoted to detective in 2006. One year later, he was hand-selected by his chief for an elite assignment to the DEA Phoenix Narcotic Task Force Team 3. And that was when I met him.

From the moment Diego and I partnered up, it was clear that our strengths complemented one another. Diego had an innate street sense. He was always *working* someone: a confidential informant, a crook—even his friends. He often juggled four cell phones at a time. The undercover role—front and center, doing all the talking—was where Diego thrived. While I loved working the street, I'd always find myself in the shadows, as I was that night, sitting at our table, taking mental note of every detail, studying and memorizing every face. I didn't want the spotlight; my work behind the scenes would speak for itself.

Diego and I had just started targeting a Phoenix-based crew of narco juniors suspected of distributing Sinaloa Cartel cocaine, meth, and large shipments of *cajeta*—high-grade Mexican marijuana—by the tractor-trailer-load throughout the Southwest.

Though we weren't planning to engage the targets that night, Diego was dressed just like a narco junior, in a black Calvin Klein button-down shirt, untucked over midnight-blue jeans, and a black-faced Movado watch and black leather Puma sneakers. I looked more like a college kid from California, in my black Hurley ball cap, plain gray T-shirt, and matching Diesel shoes.

My sons are my joy and my sadness, I repeated to myself silently. This most popular of the current narcocorridos—Roberto Tapia's "El Niño de La Tuna"—packed a lot of emotional punch in its lyrics. I could see the passion in the eyes of the crowd, singing along word for word. It seemed to me that they viewed El Chapo as some mix of Robin Hood and Al Capone.

I looked over and nodded at Diego as if I understood fully, but I really had *no* clue yet.

I was a young special agent from Kansas who'd grown up on a red-meat diet of Metallica, Tim McGraw, and George Strait, and it was a lot to take in that first night with Diego in Mariscos Navolato.

Up on the five flat-screen TVs, a big Mexican Primera División soccer match was on—Mérida was up 1–0 against Querétaro, apparently, though it meant little to me. The CD jukebox was filled with banda and ranchera, the walls covered in posters for Modelo, Tecate, Dos Equis, and Pacifico, homemade flan, upcoming norteño concerts, and handwritten signs about the *mariscos* specialties like *almeja Reyna*, a favorite clam dish from Sinaloa.

"El Chapo"? Was "Shorty" supposed to be a *menacing*-sounding nickname? How could some semiliterate kid from the tiny town of La Tuna, in the mountains of the Sierra Madre—who'd supported his family by selling oranges on the street—now be celebrated as the most famous drug lord of all time? Was Chapo really—as the urban legends and corridos had it—even more powerful than the *president* of Mexico?

Whatever the truth of El Chapo, I kept my eyes glued to the narco juniors sitting at a table near the far end of the bar. One had a fresh military-style haircut, two others fauxhawks, the fourth sporting an Arizona State University ball cap. Diego and I knew they were likely armed.

If the narco juniors went out to their cars, we'd have to follow.

Diego tossed two $20 bills on the table, winked at the waitress, and rose from his seat. Now the crew shifted in their seats, one getting to his feet, fixing the brim on his cap, pivoting on the sole of his Air Jordans like a point-guard.

Diego downed the last gulp of his Pacifico and gestured for me to do the same. The band was blaring louder now; Diego laughed, along with the entire bar, hitting the crescendo of the song:

I may be short, but I'm brave . . .

And I began to grin, too, as I slid my chair back and stood up.

The hypnotic rhythm took hold; I found myself singing with as much gusto as any of these cowboy-hat-clad traffickers:

"Yo soy El Chapo Guzmán!"

PART I

BREAKOUT

GUADALAJARA, MEXICO
May 24, 1993

THE SUDDEN BURST OF AK-47 gunfire pierced the calm of a perfect spring afternoon, unleashing panic in the parking lot of the Guadalajara Airport. Seated in the passenger seat of his white Grand Marquis, Cardinal Juan Jesús Posadas Ocampo, the Archbishop of Guadalajara, was struck fourteen times as he arrived to meet the flight of the papal nuncio. The sixty-six-year-old cardinal slumped toward the center of the vehicle, blood running down his forehead. He had died instantly. The Grand Marquis was riddled with more than thirty bullets, and his driver was among six others dead.

Who would possibly target the archbishop—one of Mexico's most beloved Catholic leaders—for a brazen daylight hit? The truth appeared to be altogether more prosaic: it was reported that Cardinal Posadas had been caught up in a shooting war between the Sinaloa and Tijuana cartels, feuding for months over the lucrative "plaza"—drug smuggling route—into Southern California. Posadas had been mistaken for the leader of the Sinaloa Cartel, Joaquín Archivaldo Guzmán Loera, a.k.a. "El Chapo," who was due to arrive at the airport parking lot in a similar white sedan at around the same time.

News footage of the Wild West–style shoot-out flashed instantly around the world as authorities and journalists scrambled to make sense of the carnage. "Helicopters buzzed overhead as police confiscated about 20 bullet-riddled automobiles, including one that contained grenades and high-powered automatic weapons," reported the *Los Angeles Times* on its front page. The daylight assassination of Cardinal Posadas rocked Mexican society to its core; President Carlos Salinas de Gortari arrived immediately to pay his condolences and calm the nation's nerves.

The airport shoot-out would prove to be a turning point in modern Latin American history: for the first time, the Mexican public truly took note of the savage nature of the nation's drug cartels. Most Mexicans had never heard of the diminutive Sinaloa capo whose alias made him sound more *comical* than lethal.

After Posada's assassination, crude drawings of Chapo's face were splashed on front pages of newspapers and magazines all across Latin America. His name appeared on TV nightly—wanted for murder and drug trafficking.

Realizing he was no longer safe even in his native Sierra Madre backcountry, or in the neighboring state of Durango, Guzmán reportedly fled to Jalisco, where he owned a ranch, then to a hotel in Mexico City, where he met with several Sinaloa Cartel lieutenants, handing over tens of millions in US currency to provide for his family while he was on the lam.

In disguise, using a passport with the name Jorge Ramos Pérez,

Chapo traveled to the south of Mexico and crossed the border into Guatemala on June 4, 1993. His plan apparently was to move stealthily, with his girlfriend and several bodyguards, then settle in El Salvador until the heat died down. It was later reported that Chapo had paid handsomely for his escape, bribing one Guatemalan military officer with $1.2 million to guarantee his safe passage south of the Mexican border.

IN MAY 1993, around the time of the Posada murder, I was fifteen hundred miles away, in my hometown of Pattonville, Kansas, diagramming an intricate pass play to my younger brother. We were Sweetness and the Punky QB—complete with regulation blue-and-orange Bears jerseys—huddling up in the front yard against a team made up of my cousins and neighbors. My sister and her friends were dressed up as cheerleaders, with homemade pom-poms, shouting from the sidelines.

My brother, Brandt, always played the Walter Payton role. I was Jim McMahon, and I was a fanatic—everyone teased me about it. Even for front-yard games, I'd have to have all the details just right, down to the white headband with the name ROZELLE, which I'd lettered with a black Magic Marker, just like the one McMahon had worn in the run-up to the 1985 Super Bowl.

None of us weighed more than a hundred pounds, but we took those front-yard games seriously, as if we really were Payton, Mc-Mahon, Singletary, Dent, and the rest of the Monsters of the Midway. In Pattonville—a town of three thousand people, fifty-two miles outside Kansas City—there wasn't much else to do besides play football and hunt. My father was a firefighter and lifelong waterfowl hunter. He'd taken me on my first duck hunt at age eight and bought me my first shotgun—a Remington 870 youth model—when I turned ten.

Everyone expected I'd become a firefighter, too—my great-grandfather, my grandfather, and three uncles had all been firemen. I'd spend hours at the fire station following my dad around, trying on his soot-stained leather fire helmet and climbing in and out of the trucks in the bay. In fifth grade, I brought home a school paper and showed my mom:

"Someday I'm going to be . . . a fireman, a policeman, or a spy detective."

But as long as I could remember, I'd really been dead set on becoming *one* thing: a cop. And not just *any* cop—a Kansas State Trooper.

I loved the State Troopers' crisp French-blue uniforms and navy felt campaign hats, and the powerful Chevrolets they got to drive. For years I had an obsession with drawing police cars. It wasn't just a hobby, either—I'd sit alone in my bedroom, working in a feverish state. I had to have all the correct colored pens and markers lined up, drawing and shading the patrol cars in precise detail: correct light bar, insignia, markings, wheels—the whole works had to be spot-on, down to the exact radio antennas. I'd have to start over even if the slightest detail looked off. I drew Ford Crown Vics and Explorers, but my favorite was the Chevy Caprice with the Corvette LT1 engine and blacked-out wheels. I'd often dream while coloring, picturing myself behind the wheel of a roaring Caprice, barreling down US Route 36 in hot pursuit of a robbery suspect . . .

Fall was my favorite time of year. Duck hunting with my dad and brother. And football. Those front-yard dreams now playing out under the bright stadium lights. Our varsity team would spend Thursday nights in a barn or some backwoods campsite, sitting around a fire and listening to that week's motivational speaker, everyone's orange helmets, with the black tiger paws on the sides, glowing in the flickering light.

Life in Pattonville revolved around those Friday-night games. All along the town's roads you'd see orange-and-black banners, and everyone would come and watch the Tigers play. I had my own pregame ritual, blasting a dose of Metallica in my headphones:

Hush little baby, don't say a word
And never mind that noise you heard

After high school, I was convinced that I'd live in the same town where my parents, grandparents, uncles, aunts, and dozens of cousins lived. I had no desire to go anyplace else. I never could have imagined leaving Pattonville. I never could have imagined a life in a smog-cloaked city of more than 26 million, built on top of the ancient Aztec capital of Tenochtitlán . . .

Mexico? If pressed—under the impatient glare of my third-period Spanish teacher—I probably could have found it on the map. But it might as well have been Madagascar.

I WAS SOON THE black sheep: the only cop in a family of firefighters. After graduating from Kansas State University with a degree in criminal justice, I'd taken the written exam for the Kansas Highway Patrol, but a statewide hiring freeze forced me in another direction. A salty old captain from the local sheriff's office offered me a job as a patrol deputy with Lincoln County, opening my first door to law enforcement.

It wasn't my dream job, but it *was* my dream ride: I was assigned a 1995 Chevrolet Caprice, complete with that powerhouse Corvette engine—the same squad car I'd been drawing and coloring in detail in my bedroom since I was ten years old. Now I got to take it home and park it overnight in the family driveway.

Every twelve-hour shift, I was assigned a sprawling twenty-by-thirty-mile zone. I had no patrol-car partner: I was just one baby-faced deputy covering a vast countryside scattered with farmhouses and a few towns. The closest deputy would be in his or her zone, just as large as mine. If we were on the opposite ends of our respective zones and needed backup, it could take thirty minutes to reach each other.

I discovered what that really meant one winter evening during my rookie year when I went to look for a six-foot-four, 260-pound suspect—name of "Beck"—who'd just gotten out of the Osawatomie State Hospital psychiatric ward. I'd dealt with Beck once already that night, after he'd been involved in a domestic disturbance in a nearby town. Just after 8 p.m., my in-car mobile data terminal beeped with a message from my sergeant: "Hogan, you've got two options: get him out of the county or take him to jail."

I knew I was on my own—the sergeant and other deputies were all handling a vehicle in the river, which meant my colleagues were twenty minutes away at a minimum. As I drove down a rural gravel road, in my headlights I caught a dark figure ambling on the shoulder. I let out a loud exhale, pulling to a stop.

Beck.

Whenever I had a feeling that things were going to get physical, I tended to leave my brown felt Stratton hat on the passenger seat. This was one of those times.

"David twenty-five," I radioed to dispatch. "I'm going to need another car."

It was the calmest way of requesting immediate backup. But I knew the truth: there wasn't another deputy within a twenty-five-mile radius.

"The Lone fuckin' Ranger," I muttered under my breath, stepping out of the Caprice. I walked toward Beck cautiously, but he continued walking away, taking me farther and farther from my squad car's headlights, and deeper and deeper into the darkness.

"Sir, I can give you a ride up to the next gas station or you can go to jail," I said, as matter-of-factly as I could. "Your choice tonight."

Beck ignored my question completely, instead picking up his pace. I half jogged, closed the distance, and quickly grabbed him around his thick bicep to put him in an arm bar. Textbook—just how I'd been taught at the academy.

But Beck was too strong to hold, and he lunged forward, trying to free his arm. I felt the icy gravel grinding beneath us as we both tried to gain footing. Beck snatched me in a bear hug, and there were quick puffs of breath in the cold night air as we locked eyes for a split second, faces separated by inches. I had zero leverage—my feet now just barely touched the ground. It was clear that Beck was setting up to body-slam me.

I knew there was no way I could outgrapple him, but I managed to rip my right arm loose and slammed my fist into his pock-marked face, then again, until a third clean right sent Beck's head snapping back and he finally loosened his grip. I planted my feet to charge, as if I were going to make a football tackle, and rammed my shoulder into Beck's gut, driving him to the ground. Down into the steep frozen ditch we barrel-rolled on top of each other, Beck trying to grab for my .45-caliber Smith & Wesson pistol, unclasping the holster snaps, nearly getting the gun free.

I finally got the mount, reached for my belt, and filled Beck's mouth and eyes with a heavy dose of pepper spray. He howled, clutching at his throat, and I managed to get him handcuffed, on his feet, and into the backseat of the Caprice.

We were halfway to the county jail before my closest backup even had a chance to respond.

It was the scariest moment of my life—until twelve years later, when I set foot in Culiacán, the notorious capital of the Mexican drug underworld. . . .

DESPITE THE DANGERS, I quickly developed a taste for the hunt. During traffic stops, I'd dig underneath seats and rummage through glove compartments in search of drugs, typically finding only half-empty nickel bags of weed and crack pipes. Then, one evening on a quiet strip of highway, I stopped a Jeep Cherokee for speeding. The vehicle sported a small Grateful Dead sticker in the rear window, and the driver was a forty-two-year-old hippie with a grease-stained white T-shirt. I knew exactly how to play this: I acted like a clueless young hillbilly cop, obtained his verbal consent to search the Jeep, and discovered three ounces of rock cocaine and a bundle of more than $13,000 in cash.

The bust made the local newspapers—it was one of the largest drug-cash seizures in the history of our county. I soon got a reputation for being a savvy and streetwise patrolman, skilled at sniffing out dope. It was a natural stepping-stone, I was sure, to reaching my goal of becoming a Kansas State Trooper.

But then a thin white envelope was waiting for me when I drove the Caprice home one night after my shift. The Highway Patrol headquarters, in Topeka, had made its final decision: despite passing the exam, I was one of more than three thousand applicants, and my number simply was never drawn.

I called my mom first to let her know about the rejection. My entire family had been waiting weeks to hear the exam results. The moment I hung up the phone, my eyes fixed on the framed photo of the Kansas Highway Patrol patch I'd had since college. I felt the walls of my bedroom closing in on me—as tight as the corridor of the county jail. Rage rising into my throat, I turned and smashed the frame against the wall, scattering the glass across the floor. Then I jumped onto my silver 2001 Harley-Davidson Softail Deuce and lost myself for five silent hours on the back roads, stopping at every dive bar along the way.

My dad was now retired from the Pattonville Fire Department and had bought the town's original firehouse—a two-story red-brick 1929 building on the corner of East Main and Parks Street—renovated it, and converted it into a pub. Pattonville's Firehouse Pub quickly became the town's busiest watering hole, famous for its hot wings, live bands, and raucous happy hours.

The pub was packed that night, a four-piece band playing onstage, when I pulled up outside the bar and met up with my old high school football buddy Fred Jenkins, now a Kansas City firefighter.

I tried to shake it off, but my anger kept simmering—another bottle of Budweiser wasn't going to calm this black mood. I leaned over and yelled at Freddie.

"Follow me."

I led him around to the back of the pub.

"What the hell you doing, man?"

"Just help me push the fuckin' bike in."

Freddie grabbed hold of the front forks and began to push while I backed my Deuce through the rear door of the bar.

I saddled up and ripped the throttle, and within seconds white smoke was billowing around the rear tire as it cut into the unfinished concrete floor.

A deafening roar—I had the loudest pipes in town—quickly drowned out the sound of the band. Thick, acrid-smelling clouds filled the bar as I held on tight to the handlebars, the backs of my legs pinched against the rear foot pegs to keep the hog steady—the ultimate burnout—then I screeched off, feeling only a slight relief.

I parked the Deuce and walked back into the bar, expecting high fives—something to lighten my mood—but everyone was pissed, especially my father.

Then some old retired fireman knocked me hard on the shoulder.

"Kid, that was some cool shit," he said, "but now my chicken wings taste like rubber."

I reached into my jeans and pulled out a wad of cash for a bunch of dinners. Then I saw my father fast approaching behind the bar.

"Let's roll," I yelled through the crowd to Freddie. "Gotta get outta here before my old man beats my ass."

I RETESTED WITH the Highway Patrol but started looking into federal law enforcement careers, too—one of my best cop buddies had told me good things about the Drug Enforcement Administration. Until then, I had never considered a career as a special agent, but I decided to take the long drive over to Chicago and attend their orientation. The process was surprisingly quick, and I was immediately categorized as "best qualified," with my past police experience and university degree. Months went by without a word, but I knew it could take more than a year before I completed the testing process. One fall morning, I was back on my Harley with a bunch of cops and firefighters for the annual US Marine Corps Toys for Tots fund-raising ride. After a long day cruising the back roads, doing a little barhopping, I let slip to Freddie's cousin, Tom, that I had applied with the DEA.

"No kidding? You know Snake?" Tom said, then called across the bar: "Snake! Get over here—this kid's applying with the DEA."

Snake swaggered over in his scuffed-up leather jacket. Headful of greasy blond shoulder-length hair, wearing a half-shaven beard and a scowl, he looked more like a full-patch outlaw biker than a DEA agent.

I hit it off with Snake right away—we downed a couple of bottles of Bud and talked about the snail-paced application process.

"Look, kid, it's a pain in the ass, I know—here's my card," Snake said, giving me his number. "Call me Monday."

Before I knew it, thanks to Snake, I found myself on a fast track through the testing process and received an invitation to the DEA Training Academy. One last blowout night at the Firehouse Pub, then I headed east, breaking free of my meticulously laid-out life in Kansas. I drove through the heavily forested grounds at Quantico—chock-full of whitetail deer so tame you could practically pet them—and entered the gates of the DEA Academy as a member of a brand-new class of basic agent trainees.

I had barely settled into life at Quantico when I got a call telling me I'd been selected as a candidate for the next Kansas Highway

Patrol class. I scarcely believed what I heard myself telling the master sergeant on the phone.

"Thanks for the invite," I said, "but I'm not leaving DEA."

By that point, I was throwing myself headlong into the DEA training.

We spent hours on the range, burning through thousands of rounds of ammunition, firing our Glock 22 .40-caliber pistols or busting our asses doing PT out near the lake's edge—sets of burpees in the icy, muddy water, followed by knuckle push-ups on the adjacent gravel road.

The heart of academy training was the practical scenarios. We called them "practicals." One afternoon during a practical, I had the "eye" on a target—an academy staff member playing the role of a drug dealer—planning an exchange with another bad guy in a remote parking lot. I parked just out of sight, grabbed my binoculars and radio, and crawled up underneath a group of pine trees.

"Trunk is open," I radioed my teammates. "Target One just placed a large black duffel bag into the back of Target Two's vehicle. They're getting ready to depart. Stand by."

Alone in my Ford Focus, I followed the second target vehicle to another set.

Time for the vehicle-extraction takedown. I still had eyes on Target Two, but none of my teammates had arrived in the parking lot. Minutes passed; I was staring at my watch, calling my team on the radio; I knew we needed to arrest the suspect now or we'd all flunk the practical.

I hit the gas and came to a skidding stop near the rear of the target vehicle, and, with my gun drawn, I rushed the driver's door.

"Police! Show me your hands! Show me your hands!"

The role player was so startled he didn't even react. I reached in through the door and grabbed him by the head—hauling him from the vehicle and throwing him face-first onto the asphalt before cuffing him.

My team passed the practical, but I caught pure hell from our instructor during the debrief. "Think you're some kind of goddamn *cowboy*, Hogan? Why didn't you wait for your teammates before initiating the arrest?"

Wait?

I held my tongue. It wasn't that easy to unwire the aggression, the street-cop instinct, honed during those years working alone as a deputy sheriff with no backup.

That tag—Cowboy—stuck with me for the final weeks of the academy.

I graduated in the top of my class and, with my whole family present, walked across the stage in a freshly pressed dark blue suit and tie to receive my gold badge from DEA Administrator Karen Tandy, then turned and shook the hand of Deputy Administrator Michele Leonhart.

"Congratulations," Michele said. "Remember, go out there and make big cases."

THE PRISON WAS his playground.

Down in Jalisco—the home of Mexico's billion-dollar tequila industry—Chapo was living like a little prince. On June 9, 1993, after successfully slipping into Guatemala, he was apprehended by the Guatemalan army at a hotel just across the border. The political heat was too intense: he couldn't bribe his way out of this jam. It was the first time his hands had felt the cold steel of handcuffs, and his first police mug shot was taken in a puffy tan prison coat.

Before long, Guzmán was aboard a military plane, taken to the Federal Social Readaptation Center No. 1, known simply as Altiplano, the maximum-security prison sixty miles outside Mexico's capital.

By now the public knew more about Chapo. The young campesino had dropped out of school and sold oranges on the streets to help support his family. Later he'd been a chauffeur—and allegedly a prodigious hit man—for Miguel Ángel Félix Gallardo, a.k.a. "El Padrino," the godfather of modern Mexican drug trafficking.

Born on the outskirts of Culiacán, Gallardo had been a motorcycle-riding Mexican Federal Judicial Police agent and a bodyguard for the governor of Sinaloa, whose political connections Gallardo used to help build his drug-trafficking organization (DTO). A business major in university, Gallardo had seen a criminal vision of the future: he united all the bickering traffickers—mostly from Sinaloa—into the first sophisticated Mexican DTO, called the Guadalajara Cartel, which would become the blueprint for all future Mexican drug-trafficking organizations.

Like Lucky Luciano at the birth of modern American organized crime, in the late 1920s, Gallardo recognized that disputed territory led to bloodshed, so he divided the nation into smuggling "plazas" and entrusted his protégé, Chapo Guzmán, with control of the lucrative Sinaloa drug trade.

While he was behind bars after his Guatemalan capture, Guzmán's drug empire continued to thrive. Chapo's brother, Arturo, was the acting boss, but Chapo himself was still clearly calling all the shots—he was now ranked as the most powerful international drug trafficker by authorities in both Mexico and the United States.

Chapo was moving *staggering* amounts of cocaine—regularly and reliably—from South America up through Central America and Mexico and into the United States. These weren't small-time muling jobs, either: Chapo's people were moving multi-ton shipments of Colombian product via boat, small planes, even jerry-rigged "narco subs"—semi-submersible submarines capable of carrying six tons of pure cocaine at a time. Chapo's methods of transport were creative—not to mention constantly evolving—and he thereby earned a reputation for getting his loads delivered intact and on time. Chapo expanded his grip to ports on Mexico's Atlantic and Pacific coasts and strong-armed control of key crossing points—not just on the US-Mexico border but also along Mexico's southern border with Guatemala.

Chapo embedded lieutenants of the Sinaloa Cartel in Colombia, Ecuador, Costa Rica, El Salvador, Guatemala, and Venezuela, giving him more flexibility to negotiate directly with traffickers within the supply chain. His criminal tentacles, versatility, and ingenuity surpassed even his more infamous predecessors, like Pablo Escobar. Headline-making seizures of Chapo's cocaine—13,000 kilograms on a fishing boat, 1,000 on a semi-submersible, 19,000 from another maritime vessel en route to Mexico from Colombia—were mere drops in the cartel's bucket, losses chalked up to the cost of doing business.

Even from behind bars, Chapo had the insight to diversify the Sinaloa Cartel's operations: where it had previously dealt strictly in cocaine, marijuana, and heroin, the cartel now expanded to the manufacture and smuggling of high-grade methamphetamine, importing the precursor chemicals from Africa, China, and India.

On November 22, 1995—and after being convicted of possession of firearms and drug trafficking and receiving a sentence of twenty years—Chapo arranged to have himself transferred from Altiplano to the maximum-security Federal Social Readaptation Center No. 2, known as Puente Grande, just outside Guadalajara.

Inside Puente Grande, Guzmán quickly built a trusted relationship with Dámaso López Núñez, a.k.a. "El Licenciado"—or simply "El Lic"—a fellow Sinaloan, from the town of El Dorado. El Lic had been a police officer at the Sinaloa Attorney General's Office before being appointed to a management position in Puente Grande prison.

Under El Lic's watch, Chapo reportedly led a life of luxury— liquor and parties, and watching his beloved fútbol matches. He was able to order special meals from a handpicked menu, and when that grew boring, there was plenty of sex. Chapo was granted regular conjugal visits with his wife, various girlfriends, and a stream of prostitutes. He even arranged to have a young woman who was serving time for armed robbery transferred to Puente Grande to further tend to his sexual needs. The woman later revealed Chapo's supposed romantic streak: "After the first time, Chapo sent to my cell a bouquet of flowers and a bottle of whiskey. I was his queen." But the reality was more tawdry: on the nights he got bored with her, it was said he passed her off among other incarcerated cartel lieutenants.

It was clear that Chapo was the true boss of the lockup. With growing fears of being extradited to the United States, he planned a brazen escape from Puente Grande.

And sure enough, just after 10 a.m. on January 19, 2001, Guzmán's electronically secured cell door opened. Lore has it that he was smuggled out in a burlap sack hidden in a laundry cart, then driven through the front gates in a van by one of the corrupt prison guards in a mode reminiscent of John Dillinger's famous jailbreaks of the 1930s.

The escape required complicity, cooperation, and bribes to various high-ranking prison officials, police, and government authorities, costing the drug lord an estimated $2.5 million. At 11:35 p.m., the prison warden was notified that Chapo's cell was empty, and chaos ensued. When news of his breakout hit the press, the Mexican government launched an unprecedented dragnet, the most extensive military manhunt the country had mounted since the era of Pancho Villa.

In Guadalajara, Mexican cops raided the house of one of Guzmán's associates, confiscating automatic weapons, drugs, phones, computers, and thousands of dollars in cash. Within days of the escape, though, it was clear that Guzmán was no longer in Jalisco. The manhunt spread, with hundreds of police officers and soldiers searching the major cities and sleepiest rural communities.

Guzmán called a meeting of all the Sinaloa Cartel lieutenants, eager to prove that he was still the top dog. A new narcocorrido swept the nation, "El Regreso del Chapo."

No hay Chapo que no sea bravo
*Así lo dice el refrán**

Chapo was not just *bravo*: he was now seen as untouchable—the narco boss that no prison could hold. Sightings were reported the length of the nation, but whenever the authorities were getting close to a capture, he could quickly vanish back into his secure

---

* Short guys are always brave
So the saying goes

redoubt in the Sierra Madre—often spending nights at the ranch where he'd been born—or back into the dense forests and marijuana fields. He was free, flaunting his power, and still running the Sinaloa Cartel with impunity.

It would be nearly thirteen years before he again came face-to-face with any honest agent of the law.

THE NEW GENERATION

PHOENIX, ARIZONA
October 5, 2008

"LAS TRES LETRAS."

I repeated the words, looking to Diego for assistance, but I got none. We were sitting in the Black Bomber on a surveillance, listening to a narcocorrido by Explosion Norteña.

Diego chewed on the end of his straw and rattled the ice cubes in his Coke cup, his brow creased like a stern teacher's.

"The *Three* Letters?"

The Black Bomber was the ideal vehicle for listening to narcocorridos—booming bass in the Bose speakers, clarity as good as at any Phoenix nightclub. When Diego first came to DEA Phoenix, he was driving that jet-black Chevrolet Suburban Z71 with heavy tint on all the windows and a tan leather interior.

The Mesa PD had seized the Suburban from a coke dealer a couple of years earlier. The owner's luxury options had made the Bomber the perfect ride for us on long surveillance operations, which included a flip-up customized video screen in the dash. We'd often kill the hours watching *Super Troopers*, parked in the shadows on a side street before a dope deal was supposed to go down.

But the Black Bomber wasn't just a rolling entertainment center on 24-inch rims; it was also ideal for raids—unlike standard cop cars, the Suburban could fit four of us in all our tactical gear. We thought of the Bomber as another team member. It was a sad day when some number-crunching bureaucrat made Diego turn her in because she had 200,000 miles on the odometer.

Diego would get pulled over in the Black Bomber by Phoenix cops all the time, simply because it had Mexican plates. He'd kept the originals from the state of Sonora, white and red with small black letters and numbers. Local cops were always looking for cars—especially tricked-out SUVs—with Mexican plates, but it allowed us to blend into any Mexican hood in Phoenix. No one would think twice about a parked Suburban with Sonora plates: behind those dark-tinted windows, Diego and I could sit on a block all night and never get burned by the bad guys.

And the narcocorridos Diego was always playing in the Black Bomber had become central to my education. Every big-time trafficker south of the Rio Grande had at least one norteño song celebrating his exploits.

You were no one in the narco world, Diego explained, until you had your own corrido. But I was still trying to decipher Las Tres Letras . . .

"Come on, brother," Diego said, laughing. "You *got* this. Shit, at this point you're more Mexican than most of the Mexicans I know . . ."

I leaned forward in the Bomber and hit the repeat button on the CD player, taking one more shot at decoding those lyrics.

"Las Tres Letras?"

Finally, Diego jabbed his index finger hard into my shoulder. "Bro, *you're* Las Tres Letras! DEA."

Las Tres Letras . . . what every drug trafficker fears the most.

DAYS AFTER DIEGO FIRST told me about El Niño de La Tuna, I'd started after-hours research in my cubicle back at the DEA office in central Phoenix.

I searched for "Joaquín GUZMÁN Loera" in our database, the Narcotics and Dangerous Drugs Information System (NADDIS). Chapo's file was endless; you could scroll down for almost an hour without reaching the end. DEA Phoenix had an open case against Guzmán, but so did dozens of other jurisdictions all across the country. I couldn't begin to fathom what I'd need to do, how many major cases I'd have to initiate, in order to be the agent entrusted with heading an investigation targeting Guzmán.

THE PRESIDENT of the United States identified Guzmán and the Sinaloa Cartel as significant foreign narcotics traffickers, pursuant to the Kingpin Act,* in 2001 and again in 2009. The US government had offered a $5 million reward for information leading to his capture, and the Mexican government had offered a reward of 60 million pesos—roughly $3.8 million.

---

* The Foreign Narcotics Kingpin Designation Act, informally known as the Kingpin Act.

Wildly divergent rumors swirled about Chapo. Some stemmed from law enforcement intel, others from street gossip—the loose chatter of informants—and some were just urban legends, embedded in the lyrics of all those underground corridos.

By one account, Chapo was considering having plastic surgery so he could never again be recognized; in another, he'd vowed to commit suicide rather than be captured alive. In May 2003, he'd been reported as living in a remote cave—a Mexican version of Osama bin Laden—but then, in June of that year, he was said to be traveling free as a bird within Mexico City. Another intel report had him hiding in Guatemala and returning to Mexico only on occasion, and in September 2004 he narrowly escaped just before a two-ton marijuana-and-weapons seizure in the Sierra Madre.

How could anyone possibly separate the facts from the fiction? Was Chapo surrounded by hundreds of heavily armed bodyguards, wearing a bulletproof vest at all times? Or was he living more simply—traveling with just two trusted associates—because he was receiving tacit protection from the Sinaloa State Police on the cartel payroll?

I DIDN'T HAVE much time to ruminate on the life and crimes of Chapo Guzmán—for more than a year, Diego and I had our hands full with thirty-one-year-old Pedro Navarro, a.k.a. "Bugsy." Bugsy's crew may have been young—in their early and mid-twenties—but they weren't small-time. Within weeks of developing my first intel on Bugsy, I received authorization to initiate an Organized Crime Drug Enforcement Task Force investigation that Diego and I titled "La Nueva Generación" (the New Generation), a Priority Target Investigation for DEA Phoenix.

Since I first saw them that smoky night in Mariscos Navolato with Diego, I'd developed a grudging respect for these narco juniors. They were savvy kids who had drug trafficking in their bloodlines—they were often the sons of heavy-hitting cartel men in Mexico—but most had gone to high school and college in the United States and Europe. That level of education, their flawless English, and their familiarity with American culture allowed them to start up their own sophisticated drug organizations. Narco juniors like Bugsy were scattered all over the Southwest, from Phoenix to San Diego.

These young men had the confidence and swagger of a new generation—and, in fact, Diego and I started referring to ourselves as La Nueva Generación as well. A mirror image of the narco juniors, we were a fresh young crop of cops with the stamina and street savvy to keep pace with young Mexican traffickers.

We'd established that Bugsy's crew was responsible for shipping ton quantities of high-grade marijuana to New York City, Baltimore, Boston, and St. Louis by tractor-trailer, FedEx, and UPS. Navarro had leased a seven-thousand-square-foot warehouse in Mesa for packaging and shipping the marijuana, which was then concealed in crate pallets disguised as scrap-metal shelving used in big-box stores. He also had several young business owners laundering his dirty millions through local Phoenix businesses. For money couriers, he used local strippers: the girls' nonstop travel allowed them to make cash pickups all around the United States. Bugsy even had a former NFL player working as a wholesale marijuana broker for the DTO.

Bugsy would often travel armed, keeping guns inside custom-made secret storage compartments, or "traps," in his Mercedes GL550 SUV. His traps were more sophisticated than the typical drug dealer's: you had to have the ignition key turned on, the left turn signal engaged, and a small plastic lever in the cupholder turned just so—three steps executed in the correct sequence—before the trap would spring open. Sometimes, wary of us on his tail, Bugsy wouldn't carry weapons in his Benz; he'd have a crew in a follow car who were armed with pistols in their own traps.

Diego and I were intercepting Bugsy's cell phones, and I enjoyed the challenge of deciphering the narco-junior code. The phrase *gangsta-up* meant they'd be traveling armed; *pool house* referred to Bugsy's four-bedroom house in Glendale; *picture of my son* was a sample of weed. Of all the lines I heard over the wire, my favorite was when we caught Bugsy openly bragging that he and his boys were living "like *Entourage* meets *The Sopranos*."

BUT BUGSY HAD a major problem: his *cajeta* supply from Sinaloa had temporarily dried up in Phoenix. One Saturday morning, Task Force Officer Nick Jones, Diego, and I were set up on surveillance outside the "pool house." We'd just "flipped the switch" and started listening to the wiretap we had on several of Bugsy's cell phones. It had taken us months of writing and rewriting federal wire affidavits each time Bugsy would drop a phone, which he did almost every week. With nothing coming across the wire, we followed Bugsy and his crew to gain any intelligence we could.

"Looks like they're packing up," Nick said over the radio. "Get ready to roll, guys." Bugsy and his crew took off in the GL550 at high speed, westbound on Interstate 8 heading out of Phoenix.

We were hardly prepared for a long road trip, but I was thankful that Nick was with us for surveillance—the whole Task Force called him "Sticky Nicky," because he'd never lose the bad guy.

Bugsy kept driving west, and every hour or so he'd dart off an exit at the last minute in an attempt to clean his tail, but we'd been following him for too long to fall for such basic countersurveillance moves. We stayed on him for close to five hours, following just far enough behind that Bugsy wouldn't notice, until finally we ended up in San Diego.

During several days of surveillance, still wearing the same clothes, we watched as Bugsy and his crew visited one stash house after another in the suburban neighborhoods of San Diego. I had the San Diego Police Department stop a Chevy Avalanche leaving one of the stash locations—the local cops seized three hundred pounds of *cajeta* in the rear bed of a truck driven by one of Bugsy's boys.

"He was planning on taking this load right back to Phoenix," I told Diego. "We need to take advantage of his drought."

"Yeah," Diego nodded. "Think I've got the perfect guy."

AFTER RETURNING TO PHOENIX, Diego and I drafted a plan: we had Diego's confidential source introduce Bugsy to a DEA undercover agent, a thirty-two-year-old Mexican American working out of the San Diego Field Division office. Like Diego, "Alex" could play the part of a narco junior perfectly.

Knowing that Bugsy was too street-smart to fall for the typical DEA "trunk flash," we lured him down to Mission Bay, where we would flash him more than a thousand pounds of marijuana stuffed inside a DEA undercover yacht equipped with cameras, recording devices, and several bronzed girls in bikinis (who were actually female undercover San Diego cops). Mixed within the thousand pounds were the same "pillows" of *cajeta* we had just seized from Bugsy's crew.

On the day of the setup, from inside our G-ride across the bay, Diego and I kept our eyes locked on the screen of the surveillance camera we'd set up in the yacht. On the boat, Bugsy was cutting into and sniffing the same pillow he'd seen at the stash pad just a week earlier.

The mirage was so convincing that Bugsy fell headlong into the trap, telling undercover agents that he needed five hundred more pounds to complete a tractor-trailer load bound for Chicago. Alex told him that the weed he'd just seen was already spoken for, en route to another buyer in LA, so Bugsy would just have to wait a week.

IN THE MEANTIME, Diego and I worked to secure indictments on the DTO and decided to rip Bugsy's money as he came to purchase the five hundred pounds.

In a TGI Fridays parking lot, Bugsy, along with his right-hand man, Tweety, met Alex, the undercover agent, and quickly flashed a quarter million in cash—rubber-banded bundles inside a chocolate-brown Gucci bag—with the expectation that he'd soon pick up his *cajeta* order at another location down the street. But before Bugsy and Tweety could get away with the cash, Diego and I pounced.

A marked San Diego police unit swooped in to a make a traffic stop on the black Ford F150. Bugsy and Tweety sped off and started tossing $10,000 chunks of cash out the truck's windows, littering miles of San Diego freeways.

We were following the chase, pulling over to recover as much cash as we could for evidence—while countless other drivers also pulled over, quickly stuffing their pockets with bunches of Bugsy's bills, then jumping back into their cars before Diego and I could stop them.

The high-speed chase continued up Interstate 5 until Bugsy and Tweety finally ran out of cash and stopped in the middle of the freeway to surrender to police, leaving behind a trail of "cash confetti," as CNN reported—$50 and $100 bills still fluttering across the highway, creating chaos during rush-hour traffic and making national headlines.

EL CANAL

PANAMA CITY, PANAMA
June 14, 2009

THE ROOFTOP HOT TUB was kidney-shaped, and the Panamanian beer was Balboa—named after the conquistador. The palms and mirrored skyline view seemed to have been laid on in thick streaks of tropical paint. Panama City gleamed like a Caribbean Dubai.

"Salud!" Diego said, hoisting a silhouette glass of Balboa. "A la Nueva Generación!"

"Salud!" I said, raising my own glass.

The New Generation had finally stepped onto the international stage.

We'd taken down Bugsy's crew that night in San Diego and Phoenix—collapsing his entire organization, seizing another thousand pounds of marijuana and more than $450,000 in assets, including Bugsy's personal yacht, a string of Mercedes-Benzes, jewelry, and bulk cash.

But with a takedown of that scope, there were bound to be

key evidentiary remnants—wide-ranging paper trails and criminal tentacles still left unexploited.

One of those loose ends happened to be in the form of Tweety's father, Gerardo, who over the past year had been selling pounds of Mexican methamphetamine to our confidential source.

Gerardo was well connected in Nogales, Mexico, and casually mentioned that he had a friend who needed some money moved. She was middle-aged, with porcelain skin, and her black curly hair was always pulled back tight in a ponytail. Aside from smuggling loads of meth and cocaine across the border from Nogales in her Toyota RAV4, Doña Guadalupe, as everyone called her, put out the word, through Gerardo, that she was actively seeking someone who could transport money. Not just a couple hundred thousand dollars, but tens of millions.

As an undercover, Diego had played dozens of roles over the years and could slip effortlessly into many personas, but he'd never posed as a money launderer before.

"This is our chance to follow some serious cash," I told him over lunch at our favorite Chinese joint in Mesa.

"Think we can pull it off?" I asked him.

I could see the wheels turning in Diego's head, contemplating ways we could win the contract from Doña Guadalupe and begin moving the numbers to which she claimed to have access.

Within the week, Diego had finagled an introduction to Doña Guadalupe, and he immediately sold her on the services of his "company." Diego seemed to be exactly the man she was looking for, but it turned out Doña Guadalupe was just a glorified go-between, a buffering layer—the first of many, as we'd soon come to find out.

And that's how we found ourselves soaking in a hot tub on the roof of a Panamanian hotel, our first time traveling abroad—so that Diego could be introduced to Doña Guadalupe's people face-to-face.

JUST A FEW HOURS before our first undercover meeting, Diego was acting as if he didn't have a care in the world. Like any good actor, he was supremely confident in his ability to negotiate his way through any business deal. But his confidence also came from meticulous preparation. We'd spent months creating our undercover legend: Diego would be playing the role of a senior executive, the director of operations for a US-based company—supposedly a covert criminal network—operating a ton-quantity drug-and-money transportation organization. Doña Guadalupe had already sold Diego to her people, including the head of a sophisticated money-brokerage-and-laundering cell led by Mercedes Chávez Villalobos and several of her associates, based in Mexico City, Guadalajara, and Bogotá, Colombia.

When Diego spoke with Mercedes, she had been aggressive, fast-talking, and extremely demanding. Diego told me she was a tough *chilanga* from Mexico City.

After doing a quick international work-up on her, I discovered that there was a warrant out for Mercedes in Amsterdam, for laundering money back in 2008. And she had connections all around the world, country-hopping almost weekly. She was always looking for a better deal, for someone trustworthy who could move hundreds of millions of dollars quickly—and strictly on a handshake.

"Do you really believe she's sitting on all this money?" The night before the sit-down, I was staring at streams of data on my

MacBook, and the dollar amounts were staggering. "She's supposedly got a hundred million in Spain. Fifty mil in Canada. Ten mil in Australia. And some two hundred million in Mexico City?"

"Look, I'm skeptical, too," Diego said, "but what other options do we have? We need to play her out to see if she can deliver."

"What we need to know," I said, "is who all of this money *really* belongs to."

"Agreed."

OUT ON THE HOTEL BALCONY, I gazed over the thin glass wall down at the city below. Mercedes was staying at one of the few luxury hotels in town that had been completely finished. So much of the Panama City skyline remained half-constructed: cranes and scaffolding and exposed girders. Brand-new buildings had been abandoned half-complete, while many of the finished ones were empty.

Panama City was the money-laundering capital of the Western Hemisphere. Banks had sprouted up on every corner like cactus along the sidewalks of Phoenix. Citibank, Chase, RBC, Bank of Montreal . . . but also lesser-known Latin American ones: Balboa Bank & Trust, Banco General, Mercantil Bank, and Centro Comercial de Los Andes . . . There was plenty of legitimate banking business, but some, like HSBC, faced criminal prosecution for "willfully failing to maintain an effective anti-money laundering program" in connection with hundreds of millions of dollars of dirty drug money belonging to Mexican cartel bosses.*

Over the months of phone-wooing, Mercedes had suggested

---

* HSBC accepted responsibility for the alleged conduct, entering into a deferred prosecution agreement with the US government.

meeting Diego face-to-face in Mexico City, but the DEA brass considered it too dangerous, and our Mexican police counterparts would never allow it. "El Canal" was perfect: Panama was known as a neutral zone for drug traffickers from all around the world to meet without threats of territorial disputes or violence. It was also geographically convenient if you wanted to meet Colombian or Mexican contacts. Many in the narco world felt at ease in this glitzy isthmus.

Eventually we wandered back to our hotel rooms. I had at least an hour of writing ahead, typing up the sixes, without which this entire Panama City operation would have no evidentiary value.*

As I slogged away on the reports, Diego sat on the edge of the bed, filling me in on the details from his recent phone conversations with Mercedes. But as the UC, Diego had to get his mind right—mingling with the locals, feeling the vibe of the city—so once he'd finished briefing me, he went down to the third-floor casino for another round of drinks. I sipped a fresh Balboa and continued banging away on the sixes. Fifteen minutes later, the hotel door opened.

"It's looking really good down there," Diego said.

"Meaning?"

"Lot of hotties." Diego smiled. "A few of them were checking me out—for real. One of them was eye-fucking me hard, brother."

"C'mon, dude, I gotta finish up this fuckin' six," I said, laughing, then Diego slid another Balboa across the desk. I took a deep breath and slapped my MacBook closed, and the two of us headed down to the third floor. Diego wasn't exaggerating. As those elevator doors opened, the casino bar was swarming with some of the

---

* Report of Investigation (DEA-6) almost always referred to by DEA agents as simply a "six."

most beautiful women I'd ever seen—some in slit miniskirts, tube tops, stiletto heels, and tight jeans showcasing the work of some of the top Colombian plastic surgeons.

It took a few minutes of Spanish small talk before I realized these women were all high-dollar Colombian prostitutes on "work visas" from Medellín, Cali, and Bogotá. Diego shrugged, and we decided to hang out with the girls anyway, dancing as a live band played, even though I had no idea what I was doing—the merengue steps were easy enough to fake, but with the sophisticated swirling salsa moves, I had to let my *colombiana* lead. Then we all hopped in a cab and headed out to one of the city's hottest nightclubs. A few more drinks, a little more dancing. Then another club . . .

Diego and I made it back to our rooms just in time to get three hours of sleep before the big meet. But Diego's mind was right now: he was ready to negotiate with some of the Sinaloa Cartel's most powerful money brokers. This became the typical pattern for our first night in any foreign country: we'd tear it up until nearly dawn, taking in the nightlife like the locals and getting a firsthand understanding of the streets, which would prove invaluable when we entered UC meetings.

When I was on the verge of sleep, I caught a flash of an infamous face on my hotel room TV. In Spanish I heard that, for the first time, *Forbes* had listed Joaquín "El Chapo" Guzmán as a billionaire, one of the richest and most powerful "businessmen" in the world.

WE HAD SELECTED a popular high-end steakhouse called La Rosita—located just inside the front door of a luxury shopping mall—for the next day's undercover meet with Mercedes Chávez Villalobos.

The plan was this: Diego and Mercedes would sit at an out-door table so I could keep my eyes on my partner throughout the meeting from inside the cab of a Toyota Hilux pickup, the G-ride that belonged to one of the DEA agents permanently stationed in Panama.

Neither Diego nor I could carry: Panamanian law wouldn't al-low us to bring our handguns into the country. But Diego was armed with one high-tech gadget: a secret key-fob camera that looked like an ordinary car key remote but was capable of dis-creetly recording hours of audio and video.

Diego was dressed in a well-tailored three-button dark gray suit, a white shirt, and a solid maroon tie pulled so tight it made the bottom of his neck puff out against his collar.

"Kill it, baby," I said, leaning over, hugging him. Diego nod-ded, mouth drawn tight as if he were already running scenarios in his head.

I set up the G-ride in the busy parking lot as close as I could to watch Diego enter the restaurant, discreetly parked, but with a perfect line of sight to the terrace tables.

But after two minutes, there was still no sign of Diego.

Three minutes passed. Then five. Then seven. I still couldn't see him on the terrace. I thumb-typed a text in our prearranged code, in case they checked his phone: innocuous Mexican slang for "What's happening, dude?"

"K onda, güey?"

No reply from Diego.

"K onda?"

My leg began twitching nervously.

I kept hitting resend on the BlackBerry.

Nothing.

I felt sweat drenching the front of my shirt.

This was the worst scenario for an undercover meet: we had no backup agents inside the restaurant with eyes on the UC, and no armed Panamanian counterparts watching our backs.

I couldn't sit for another second. I bolted from the Toyota and headed straight for the entrance of La Rosita.

What if Mercedes had switched up locations at the last minute?

What if her people had snatched Diego to pat him down, make sure he wasn't a cop?

In the restaurant, the hostess smiled and, in heavily accented English, said, "You have a reservation, sir?"

I was so focused, scanning for Diego's gray suit at the restaurant tables, that I barely heard myself answer.

"No, I'm meeting a friend," I said. "He's already seated."

I scanned every table hard but didn't see him anywhere.

Fuck! Had they grabbed him already?

I started to feel everyone's eyes locking on me as I frantically walked through the tables.

I hope to hell we're not compromised.

Where is he, for fuck's sake?

I had nowhere to go. I spun in a circle in the center of the restaurant, the walls becoming a blur. I quickly grabbed a busboy by the shoulder.

"El baño?" I asked, and no sooner had the kid gestured to the left than I saw that I was standing right next to Diego—in fact, I was literally looking down on the crown of my partner's head.

Diego was in an intense but muted conversation with Mercedes. And not only Mercedes, but two older Mexican-looking males. They were heavy hitters, I could tell. One appeared to be wearing a pistol, bulging behind the flap of his tan blazer.

Three targets? The meet was only supposed to be with Mercedes. I knew that Diego would be trying to hold his own, with no backup for his story, but even at a quick glance, I sensed that the sit-down had turned tense. Mercedes and the two henchmen had hard gazes; they weren't buying Diego's story.

Before anyone noticed me looking, I darted for the bathroom. A single trickle of sweat ran from my chest down to my navel. I could hear myself breathing loudly. Right before I reached the bathroom, I noticed a steak knife on a table ready to be cleared.

Could I grab it without being seen? There was no other option. I needed a weapon and had to take the chance.

As quickly as I could, I snatched up the knife, placed it flush against my wrist, and slipped it into my pocket.

In the bathroom, I turned on the sink and splashed cold water on my face, attempting to calm my nerves, hoping one of the bad guys wouldn't stroll in suddenly to take a piss.

What the hell can I do if they plan on kidnapping Diego? What if this meet is all a setup to take him as human collateral?

The door suddenly swung open—I straightened up, my face still dripping with cold water, but it was just a regular restaurant patron. I knew one thing: it was crucial to get photographs of Mercedes and the two heavies so I could identify them if they took Diego by gunpoint. It would also be critical for future indictments, and I couldn't rely on the key fob Diego was carrying.

I had the steak knife ready in one pocket; in the other, I had a small Canon digital camera, which I flipped on, to video mode.

Keep the camera steady in your hand. Don't make eye contact. They won't see it's on—just stroll by naturally . . .

I walked slowly past Diego, unable to aim the Canon's lens, just hoping I'd capture the faces of everyone at the table as I walked toward the door. I knew I couldn't hang out in the restaurant alone, so I found a discreet place outside where I could watch Diego through the windows of the front door. I sat there, my hands trembling as I waited for Diego to exit.

AFTER ANOTHER HOUR, Diego got up from the table, shook everyone's hands, and gave the half-hug—Mexican style—to all three, then walked out of the restaurant.

I followed him on foot as he walked on into the mall, staying thirty yards behind, making sure we weren't being followed by any of Mercedes's people.

Finally, I looked back over my shoulder three times and met up with him in a back parking lot. We were clean. We jumped in the cab of the Hilux and sped off.

Diego was silent for a long time, staring out the window and trying to make sense of what had just happened. His expression was trancelike.

"You all right, brother?" I reached over and grabbed him by the shoulder, attempting to shake him back to reality.

"What?"

"Bro, you cool?"

"That was so fuckin' intense," Diego said at last. "A straight-up interrogation. She kept hitting me with question after question. 'Who's your company? Who do you work with?'"

"How'd you play it?"

"Just started making up shit, story after story—how we're moving millions in tractor-trailers, our fleet of private aircraft. Ships. Told them we transport coke—by the tons."

"And?"

Diego grinned.

"She bought it, man!" he shouted. "She fuckin' bought it! I had all three of them eating out of the palm of my hand."

"Outstanding! Did she say whose money it is?"

"Yeah, it's his," Diego said.

"His?"

"She said it's *his*," Diego repeated.

Diego went quiet, smiling.

"His?" I asked again.

"Chapo."

"Chapo."

"Yes. She said, 'It's all Chapo's money.'"

TEAM AMERICA

PHOENIX, ARIZONA
July 1, 2010

I FELT LIKE a millionaire. And I *was* one—for a few hours, at least. I'd been entrusted with $1.2 million in laundered drug proceeds, freshly withdrawn from our undercover account at a local bank in Phoenix. Along with three other Task Force officers, I painstakingly counted and recounted that million in cash and stuffed the bundles into two white FedEx boxes.

The money seemed fake. It was a sensation I'd become accustomed to in the past year: anytime I handled US currency used in our undercover operations, I felt like I was thumbing through Monopoly money. A good cop is able to dissociate from the awe of the green. Those fat stacks of cash on our big conference table were just another tool of the undercover trade.

Counting the bills, I thought back to four months earlier, when Diego and I had made our first pickup. After nearly a year of nothing but big talk—her "hundred-million-dollar contracts all over globe"—Mercedes finally came through: she had a much smaller cash drop of $109,000, delivered, fittingly, in a laundry detergent bucket to Diego and my undercover teammate in a Home Depot parking lot just south of Los Angeles. That very afternoon—and following the instructions laid out precisely by Mercedes—Diego and I had run the stack to the bank, then wired the money to an account at Deutsche Bank in New York. From there, the money was transferred to an account at a corresponding bank in Mexico. Back at the office, Diego sent a photo of the wire confirmation to Mercedes over his BlackBerry and put his feet up on his desk.

"We're big-time now, dude," I said with a sarcastic laugh. It was a modest start, considering some of the huge figures Mercedes had been throwing around, but soon Diego and I were inundated with money-pickup requests. Mercedes set up back-to-back drops in New York City, black duffel bags stuffed with dirty money: $199,254 one day, $543,972 the next, and then $560,048. Always with the same wiring instructions—to a Deutsche Bank in New York.

A lot of the money couriers hardly looked the part. One time we flew to New York and followed a couple in their seventies who had parked their RV, with California plates, on a side street off Times Square, then marched two suitcases full of cash to our undercover in the shadows of the billboards.

Then it was up to Vancouver, Canada, for a pickup up of more than $800,000. The Canadian dollars had to be quickly converted to US currency before we could send the wire to Mercedes. In less than a month, we'd laundered more than $2.2 million of Chapo's money for Mercedes.

The money-laundering aspect of the investigation was authorized under an official Attorney General Exempt Operation (AGEO). An AGEO allowed federal agents to follow the money and further exploit their investigations, ultimately leading to the dismantlement of an entire drug-trafficking organization, as opposed to arresting a couple of low-level money couriers. It took me months of writing justifications to become authorized to create fictitious shell companies and open undercover bank accounts.

We'd flipped so many members of Bugsy's crew, getting them to cooperate, that our assistant United States attorney, before every proffer, would say to the defendants, "Now that you've seen the evidence we have against you, how would you like to come on over and join Team America?"

We'd dubbed our new case "Operation Team America." By June 2010, it was obvious that Mercedes had her hands full, and in the midst of all the cross-country cash pickups, she introduced Diego to Ricardo Robles, a thirty-four-year-old Mexican with a youthful face and thick black hair. Ricardo was a powerhouse money broker who'd grown up in the lucrative world of Mexican *casas de cambio*—money exchanges—even owning a few himself.

Diego and I quickly learned that all of the pickup contracts had come from Ricardo. Mercedes was just another protective layer, another buffer, shielding the true bosses while still taking her cut.

Over the weeks, Diego and Ricardo formed a tight bond. Finally, Ricardo asked for a face-to-face meet at Diego's Phoenix office. There was just one minor issue: we didn't *have* one.

Ricardo was flying in that afternoon. We arranged to have him picked up curbside at Phoenix Sky Harbor International Airport in a silver Mercedes CL 63 AMG. Our undercover teammate, driving the Mercedes, looked the part of a young narco. Following in a black Cadillac Escalade with twenty-two-inch black custom wheels, we used more undercover agents from the Task Force, posing as Diego's own personal security group.

As Ricardo drove in from the airport, we were still making the final arrangements in a luxurious high-rise office suite we'd rented. It was a gorgeous 1,200-square-foot space overlooking downtown Phoenix.

"Shit, we've got a major problem," I said to Diego as we were walking around the suite admiring the view.

"The place doesn't look lived in," Diego agreed. "Looks like we moved in five minutes ago."

I ran to the elevator, rode down to the street, jumped into my G-ride, and drove to my house. There, I grabbed some framed art off my living room walls, a few houseplants, sculptures, and trinkets I had collected from my travels. Meanwhile, Diego at the last moment set a framed photo of his kids on top of the desk. With my stuff plus his, the illusion was complete: Diego, wearing a silver Armani suit, sat back in his tall leather swivel chair, looking every bit the sleazy corporate executive.

Our teammates radioed to me that Ricardo had arrived and was coming up in the elevator. Diego quickly tightened his tie while I gave him a heavy pat on the back and rushed out the door.

We were charging at least seven percent on every money pickup, a standard commission. We would then take the commission and set it aside as Trafficker Directed Funds (TDF), to be used to rent the office space, buy the latest MacBooks and sophisticated recording devices—hidden inside expensive-looking wristwatches—along with the iridescent Armani suit for Diego.

The trust had already been established through laundering a couple million in drug money, and now it was time for Ricardo and Diego to talk about the other end of the equation: a two-ton cocaine transportation contract, moving product from Ecuador to Los Angeles.

"He wants to introduce me to Chapo's people," Diego told me as we debriefed the meeting over cups of coffee.

It had become obvious that, just like Doña Guadalupe and Mercedes, Ricardo was yet *another* buffer—another broker in the middle. And we knew there would likely be several more layers to sift through before we got to the top.

But before the introductions to Chapo's people could be made, there was a final test. Ricardo had several money pickups in Vancouver, Canada. But this time he wanted that cash—the $1.2 million—delivered directly to Mexico in *bulk*.

DIEGO HAD ALREADY FLOWN to Mexico City to coordinate the operation with DEA agents there, as well as trusted members of the Mexican Federal Police (PF). Alone, I took the FedEx boxes to a Learjet—used by the DEA solely for undercover ops—through a private hangar at Sky Harbor International Airport. As we rose through the clouds, I felt like nodding off, but I didn't dare take my eyes off the cash-stuffed pair of boxes. My eyes stayed glued to them the entire flight as if they were my newborn twins.

The pilots flew me down to Toluca, outside Distrito Federal, where I was picked up by Mexico City–based DEA agent Kenny McKenzie, driving a white armored Ford Expedition. I tucked the FedEx boxes in the backseat, glancing warily around me.

Shouldn't we be covered by another armed agent?

I felt twitchy, but kept my thoughts to myself as we pulled away from the airport. It was an hour-long drive over the small mountain range into Mexico City, a route on which there was an ever-present risk of a carjacking.

We drove directly to an underground parking garage in a middle-class area called Satélite, on the north side of the city. I was relieved to see Diego, a second Mexico City DEA agent, and two plainclothes Mexican Federal Police when we arrived.

The PF had provided the "drop vehicle," a white Chevy Tornado pickup that looked to me a bit like a mini El Camino. It was a seized drug smuggler's truck, complete with a hidden stash compartment—a simple hollow cavity underneath the bed—nowhere near as sophisticated as the type Bugsy and his crew of narco juniors used in Phoenix. The deep trap—clearly designed for moving bulkier contraband like compressed bales of weed or bricks of cocaine—was accessed from behind the rear bumper, and the open space underneath ran the full length of the truck bed. Diego and I tied the FedEx boxes together, then used the remaining rope to secure them to the outside of the trapdoor so that they didn't slide the full length of the trap and become invisible to our targets.

Our Mexican counterparts also slapped a tiny GPS tracker on the truck so we could follow it to wherever the targets would unload the money, in hopes of pinpointing yet another location, another piece of the puzzle, more targets to ID, and another chance to follow the money. I kept repeating the mantra, *Exploit, exploit, exploit*, which had been drummed into my head back at the DEA Academy.

Mexican PF was doing us a huge favor by allowing us to drop $1.2 million and let it walk, but they felt they should remain on the perimeter and keep their hands off the actual money. As a result, none of the Federal Police wanted to touch the Chevy pickup, let alone drive it.

Now that the truck was loaded, Diego was in phone contact with the targets, and they agreed to pick up the vehicle on the upper-level parking lot of another mall called Plaza Satélite. Diego and the other Mexico City agent drove the armored Ford Expedition ahead while Kenny and I hopped into the shitty Tornado, with its manual transmission. Kenny drove, following the Expedition out of the parking garage and out onto the busy street, heading north. My mind was still racing:

Our security setup is worthless: more than a million in cash, and we've got a grand total of four US agents? Only two of us have Glocks—useless if we get jacked by some assholes with AKs . . .

If the operation went south, there was no way this truck would get us out of harm's way. The tiny Chevy took a day and a half to get up to 40 miles per hour. We lurched and jolted along in traffic as Kenny grinded the stick shift into gear.

It was blazing hot in the cab, and the A/C was busted. Cars, motorcycles, and trucks were buzzing, honking, zigzagging. This was the wild, chaotic traffic for which Mexico City is famous— and which I would come to know well in the years ahead. Kenny seemed to be hitting every possible pothole and red light on the route, too.

By far the biggest security risk was the local cops. Too many Federal Police knew about the operation for my liking. And if just *one* of these PF guys was dirty, he could easily call up one of his friends and ambush us, and they'd split the proceeds fifty-fifty.

The Chevy kept jumping forward while I continued talking to Diego in the Expedition on my Nextel. All of a sudden, the Expedition pulled over to the side of the road and the driver opened up the door and began projectile-vomiting onto the street. He had eaten at some roadside carnitas stand an hour earlier.

By the time we reached Plaza Satélite, one of the largest shopping malls in the city, I began to think there must be something wrong—how could a popular shopping center be so desolate?

Diego and I had no idea whether the targets were waiting at the location. We were twenty-five minutes early, but the crooks could be early, too. Kenny drove to the upper parking lot on the north side and pulled the truck in alongside a few stray cars. I sat waiting for the clear signal from surveillance to get out. We'd leave the truck right there with the keys in the ignition for the prearranged drop.

I was about to bail from the passenger side when I looked up and saw a Mexican guy, early thirties, five foot nine, muscular build, walking slowly in front of the truck. I felt my gut clench—were the crooks here already?

The guy was wearing a black-collared button-down shirt with a dark gray jacket and dark blue jeans. His eyes were a piercing brown. There was a blade wound running straight down from his left eye, a good two inches long, like he'd been disfigured by an acid teardrop.

It wasn't just the scar that was unnerving. As a street cop, you develop a sense for these things. I studied the walk: he looked to be packing on the right side of his waistband. The guy had the unmistakable gait and look of an enforcer. He walked past the pickup, looking back one more time with menace.

I turned to Kenny. "Who's that?"

"No idea, bro."

"Kenny, we need to get the fuck outta here before we get shot."

We both swung open the doors of the Tornado at the same time.

I couldn't spend another second sitting on the million-dollar bull's-eye.

———

THE TORNADO DROP was unprecedented—no federal law enforcement agency had ever delivered this kind of cash and let it walk, certainly not on the streets of Mexico City.

Diego and I were now seen by Chapo's people as fast-moving international players: we could deliver more than a million bucks, quickly—a mere forty-eight hours after picking up the bundles of bills nearly three thousand miles and two international borders away.

There was no way Ricardo would suspect that he was dealing directly with cops—let alone the DEA. Ricardo told Diego that the money was headed south to purchase a major consignment of cocaine bound for the States. It was all happening so fast that Diego and I struggled to keep pace with the logistics. We were spending more time in the air and at hotels than in the Phoenix Task Force office. We'd be on a jet to the Caribbean one week, then back at our desks in Phoenix the next, and back out on a plane the following week for yet another tropical meet.

Finding neutral countries in which to meet bad guys was becoming increasingly challenging, so I ordered a five-foot-long world map and pinned it up on the office wall. For fun, Diego and I closed our eyes and pointed a blind finger at possible locations for the next undercover meet. His finger landed on Iceland, mine somewhere in the middle of the Pacific Ocean.

Diego narrowed his focus to the isthmus of Central America, north of Panama.

"San José," he said. "Let's set the next meet in Costa Rica."

"Costa Rica sounds right," I said.

Costa Rica, like Panama, was considered neutral ground for

narcos. *Más tranquilo*, and far less risky than having a sit-down in Mexico or Colombia.

BY THE NEXT DAY, Diego was sitting across the table from two of Chapo's operators and Ricardo at an outdoor restaurant in the heart of Costa Rica's capital.

This time, unlike in Panama, I had eyes on him, parked across the street inside a rented black Toyota Land Cruiser. If Diego had felt cornered during his meet with Mercedes in Panama, this time he took the upper hand, leaning in forcefully, doing almost all the talking, pressing them with questions—the bulk cash delivery had given him the power of street credibility.

Diego asked—no, *demanded*—to know who all the coke and cash belonged to, who really was the *jefe*, before he'd set any wheels in motion.

It took him about fifteen minutes, but finally one of Ricardo's men reluctantly coughed up the name of the man they'd previously been calling El Señor.

"Carlos Torres-Ramos."

The name didn't ring any bells for Diego or me.

Jetting back to the Phoenix Task Force, I quickly began working up Carlos within the DEA's databases and found his record: Carlos Torres-Ramos had so far flown under DEA's radar, but he did have a notable criminal history. Confidential informants reported that Carlos was known for moving massive loads of cocaine by the ton from Colombia, Ecuador, and Peru. I studied the black-and-white photo. He stood six feet tall, with receding black hair, a neatly trimmed black goatee, and dark eyes that made him look almost like a professor. But there was another detail that immediately leapt out at me.

"You're not going to believe this," I said, still staring at the computer screen. "Diego, get over here."

I showed Diego the link: Carlos's daughter Jasmine Elena Torres-Leon was married to Jesús Alfredo Guzmán Salazar, one of Chapo's most trusted sons.

"Holy shit," Diego said softly. "Carlos and Chapo are *consuegros.*"

The word had no precise English equivalent—"co-fathers-in-law"—and was an important connection between two Mexican families, especially in the world of Sinaloan narcos.

We had thought Carlos was a big-time player, but we never imagined he was this big.

Diego began speaking to Carlos about transportation arrangements directly over the phone, then via BlackBerry Messenger—Carlos considered BlackBerry the most secure mode of communicating. Though they'd still never met—Diego in Phoenix, Carlos in Sinaloa—the two were establishing trust.

"*Cero-cincuenta,*" Diego said, smiling, finishing up a text session with Carlos. "Think I *got* this dude."

"*Cero-cincuenta?*"

"He just assigned me a number—like he considers me part of his organization. He calls me *cero cincuenta.*"

Diego was now "050" and part of Carlos's secret code list. All of Carlos's most trusted men were designated by a number. Locations were digitized, too: 039 represented Canada; 023 was Mexico City; 040 was Ecuador.

Carlos even sent Diego the equation his organization used to decode phone numbers when they'd send them via text. Sophisticated traffickers never give out phone numbers openly, so Diego would have to multiply every digit via the equation to figure out Carlos's new cell phone number.

———

THE MONEY PICKUPS kept flowing in from Canada, now by the millions, all going toward Carlos's purchase of the two-ton load of cocaine down in Ecuador. Of course, Diego and I weren't working for free; Diego knew the rules of the narco game and convinced Carlos to give him a deposit to cover the initial costs of transportation. Carlos agreed, and the next day he had a total of $3 million delivered to various pickup locations in Montreal and New York.

Three million in cash: as good as a seizure—and it had the added value of not burning our undercover investigation. With the money deposited in our TDF bank account, Diego and I jumped on the next plane to Ecuador to begin preparing to take delivery of the two tons of cocaine.

Once we'd arrived, Diego had a quick undercover meeting with several of Carlos's men at one of Guayaquil's upscale steakhouses. I sat at a table across the restaurant, in the shadows. This time I had a small backup army: a team of plainclothes Ecuador National Police. This was the DEA's most trusted Sensitive Investigation Unit in-country; every officer had been personally trained at Quantico in counter-narcotics operations. The plainclothesmen were spread all over the restaurant—inside and out—watching every move of Carlos's men.

ONCE DIEGO FINISHED the meeting, the cops in unmarked cars followed the men to the outskirts of the city—the crooks made a brief stop at a shop to buy brown packing tape—then to a nondescript *finca* (small farm). Covertly surveilling the *finca*, the cops were able to obtain the license plate of a white delivery truck parked outside.

Classic Quantico scenario. I remembered from my days of practicals at the academy. The events taking place were standard drug-trafficking methods.

The Ecuadorian cops sat on the truck the entire night and watched it pull away from the *finca* the next morning, the rear end loaded with bright yellow salt bags. Diego and I instructed the cops to set up a seemingly routine roadside checkpoint, and the truck drove right into it. As soon as the driver saw the marked police cars with flashing lights, he screeched to a stop, bailed out, and sprinted across a field. The police quickly chased him down and put him in cuffs. The cops searched the back of the truck and found 2,513-kilogram bricks of cocaine, stamped with the numbers 777—wrapped in that brown packing tape—tossed into seventy yellow salt bags.

Diego quickly passed the news to Carlos, via BlackBerry, that the load had been seized by the local police, but the boss didn't flinch. He'd lost two thousand keys to a random roadblock, but it was just the cost of doing business. He wasted no time in asking if Diego was ready to take delivery of more cocaine.

"You believe this guy?" I asked Diego. "Ice in his veins. Just lost a load with a street value of nearly sixty-three million and he wants to trust us with more."

Diego responded to Carlos's text immediately:

"Estamos listos. A sus ordenes."

We're ready. Awaiting your orders.

OVER THE NEXT SEVERAL WEEKS, Carlos's crew in Ecuador delivered more than eight hundred kilos of cocaine to undercover Ecuadorian cops posing as Diego's workers, triggering a global takedown of the Carlos Torres-Ramos drug-trafficking organization.

The whole house of cards came tumbling down in just a matter of hours—Carlos, Ricardo, Mercedes, Doña Guadalupe, and fifty-one other defendants spanning from Canada to Colombia. We also directly seized more than $6.3 million and 6.8 tons of cocaine.

It took Diego and me *months* to recover from the follow-up work generated by our massive takedown.

AS SOON AS THINGS settled down at the Task Force office, we were eager to get back on the hunt. But this time, we were left with only one place to go. We laid out the chart of the Sinaloa Cartel hierarchy and saw only one target name higher than Carlos. It was that pudgy-faced man with the black mustache in the photo wearing a black tactical vest and plain baseball hat, lightly gripping an automatic rifle slung across his chest.

Joaquín Archivaldo Guzmán Loera—El Chapo himself.

PART II

LA FRONTERA

IN JANUARY 2011, I put in for an open position at the DEA Mexico City Country Office, long considered one of the most elite foreign postings for US federal drug enforcement agents targeting Mexican cartels. If I hoped to successfully target Chapo Guzmán, I knew I'd need to work—and *live*—permanently south of the border. Violence was soaring in Mexico: more than 13,000 people were dead as a result of Chapo's gunmen and other cartels—notably the ex–Mexican Special Forces known as Los Zetas—battling for key smuggling turf along the US border.

Several months after we took down the Carlos Torres-Ramos organization, Diego and I began conducting our own deconfliction on Guzmán. Surely there had to be someone—*some* federal team or task force—targeting the world's most wanted narcotics kingpin. Diego and I ran through the various scenarios as we walked out of the US Attorney's Office in downtown Phoenix. There had to be agents in every federal law enforcement agency who had a bead on Chapo. We needed to find those agents, put our intel together, and begin coordinating.

I was expecting to discover a hidden world of US-agency-led Chapo task forces, secret war rooms all lining up to get their shots in—but after days of conducting deconfliction checks, Diego and I kept drawing blanks.

Who was targeting Chapo?

The shocking answer was: no one. There was no dedicated team. No elite task force. Not a single federal agent with a substantial case on his whereabouts.

Among the stacks of closed-case files, stale intelligence—not to mention the tens of millions spent each year on the "war on drugs"—Diego and I couldn't find one lawman on either side of the border who was actively pursuing the man personally responsible for controlling more than half the global drug trade.

THEN, ON FEBRUARY 15, 2011, Jaime Zapata and Víctor Ávila, two US Department of Homeland Security Investigations ("HSI") special agents on assignment in Mexico City, were ambushed in the northern state of San Luis Potosí by masked members of the Zetas Cartel. One Zetas vehicle passed the agents' armored Suburban, firing automatic rifles and ramming them off the road. The Zetas gunmen then pulled open the driver's side door and tried to drag Zapata out, but he fought back, trying to reason with the Zetas as they surrounded the vehicle. "We're Americans! We're diplomats!" The response was a hail of automatic gunfire. Zapata was killed at the wheel and Ávila badly wounded.

The murder of Special Agent Zapata threw my life into sudden turmoil. I'd already been selected for the position in Mexico City—but now I had my young family to think about, too. Was it safe to move my wife and our young sons south of the border? The majority of DEA agents wouldn't even consider putting in for a job in Mexico, due to the fear of being kidnapped or killed.

"Jesus, with Zapata getting murdered, I'm on the fence now," I told Diego. "We're happy and safe here in Phoenix, but, I dunno—this feels like the next step." We were camped out at a table at Mariscos Navolato, ties loosened, drinking a couple of Pacificos after a long day of organizing evidence for the Team America prosecutions. I was practically hoarse from talking to Diego over the blaring banda playing on the stage in front of our table.

"You know what you're getting yourself into," Diego said. "At the end of the day, you gotta do what's right for you and your family."

The next morning, I sat down with my wife and laid it all out for her. There was nothing to hide; all the risks were evident. I had been prepping her for months, but the danger of life in Mexico still weighed heavily on my mind.

"What's your gut telling you?" she asked. "I'll support you, whatever you decide."

I sat silent at the kitchen counter for a long time.

"Go," I said finally. "My gut's telling me go. Take the assignment in Mexico."

Looking back to my sheltered life in Kansas, I would never have fathomed those words. But every time I'd been faced with a life-changing decision, it felt uncomfortable, and I knew this was just another one of those moments. I paused and took a deep breath; my worries about any danger ahead began to lift: yes, it was natural career progression, after all—just furthering the investigations that Diego and I had started all those years ago.

Then it was off to six months of Spanish immersion at the DEA's language school in Southern California and, later, several more weeks of intensive training back at Quantico.

Federal agents assigned to work in high-risk foreign posts were drilled in "personnel recovery" techniques: evasive driving maneuvers, including how to take over a moving car when your partner gets killed at the wheel and how to saw through plastic handcuffs using a piece of nylon string. This was followed by specialized training in handling heavy armored vehicles, which had become mandatory after the murder of Special Agent Zapata.

IN FEBRUARY 2012, while I was away at language school, Diego had flipped a member of Chapo's inner circle traveling to the United States.

Diego called me—he was walking fast somewhere down the street, out in the wind—and he sounded breathless.

"Yo, I got his BlackBerry PIN."

"Got whose PIN?"

"C."

As always, we avoided saying the name Chapo whenever possible.

"C's BlackBerry?"

"Yup. I have *his* personal PIN."

"Holy fuck. Where's it pinging?"

"Cabo," Diego said.

"He's in Cabo San Lucas?"

"Yep—but here's the thing," Diego said, frustrated. "No one fuckin' believes me. They keep telling me it can't possibly be his number. But I'm telling you: it's him, brother."

Diego had passed the PIN to DEA Mexico City, which began its standard deconfliction. Several hours later, Diego heard back from a special agent in Mexico who told him the FBI in New York had thousands of wire intercepts with that same PIN. But they were oblivious and had no idea it was actually Chapo using it.

"Shocking," I said. "The Feebs have been secretly targeting Chapo and don't even know which goddamn phone he's using."

The DEA agent in Mexico City told Diego they were already preparing an operation with the Mexican Federal Police and had brusquely pushed Diego to the side.

"They're not going to let me in on the op," Diego said. "Shit, I should be in Cabo running this thing." I could tell Diego was feeling the strain of not having me by his side to iron out my own DEA agents in Mexico. I felt equally helpless sitting there in my Spanish immersion class, but I knew there was no stopping this runaway train—not with the Mexico City office having already involved the Mexican Federal Police.

CABO SAN LUCAS, at the tip of the Baja Peninsula, was long considered one of the safest locations in Mexico and a favorite vacation spot among Hollywood stars and thousands of American tourists. US Secretary of State Hillary Clinton was in town at the same time as Chapo—at the Barceló Los Cabos Palace Deluxe, attending a G20 foreign ministers meeting, during which she signed the United States–Mexico Transboundary Agreement.

Chapo clearly felt safe, even untouchable. DEA Mexico City put together a rapid operation that included three hundred Mexican Federal Police and moved them all up to Cabo overnight.

But the mission turned into a debacle. The takedown team launched on an upscale neighborhood of beachfront mansions, raided twelve houses . . . and came away with nothing. All they managed to do was roust a bunch of wealthy American retirees, vacationers, and well-heeled Mexican families, pissing off the entire neighborhood.

After the first failure, the Federal Police, fed up with taking grief from the community, sent most of its personnel home. DEA coordinated a second capture op, but now they didn't have enough manpower—only thirty PF officers. Nevertheless, they narrowed the pinging of Chapo's phone down to one of three beautiful beachfront mansions in a cul-de-sac right outside Cabo. As they hit the first two houses, Chapo was waiting in the third and watching it all unfold. He had no heavy security detail—the only people with him were his most trusted bodyguard, who went by "Picudo," a Cessna pilot, his cook, a gardener, and a girlfriend.

As the DEA and the PF descended on the cul-de-sac, Guzmán and Picudo slid out the back door and ran up the coast, narrowly escaping the dragnet. The two men somehow made it all the way up to La Paz and then were picked up on a clandestine airstrip— likely by Chapo's favorite pilot, Araña—and flown by Cessna back to the mountains.

After the debacle, the Associated Press reported,

> Mexican authorities nearly captured the man the U.S. calls the world's most powerful drug lord, who like Osama bin Laden, has apparently been hiding in plain sight. Federal police nearly nabbed Joaquín "El Chapo" Guzmán in a coastal mansion in Los Cabos three weeks ago, barely a day after U.S. Secretary of State Hillary Clinton met with dozens of other foreign ministers in the same southern Baja peninsula resort town.[*]

---

[*] Associated Press, March 12, 2012.

Among the people of Mexico, the raid immediately became a running joke: The Federal Police could muster a small army to capture Chapo in his mansion, but they forgot to cover the back door.

No one on the ground from DEA Mexico had a clue how big this Cabo opportunity had been. There were technological failures in the first raid, and a poorly coordinated effort in the second. The Mexicans may not have had enough people to cover the back door, true, but where the hell were the Americans? There weren't any DEA agents covering the back door, either.

A narcocorrido instantly hit the streets, recorded by Calibre 50. "Se Quedaron a Tres Pasos" ("They Stayed Three Steps Behind") turned the escape into another Dillinger-like legend, claiming that Chapo had gone on vacation in Los Cabos and then "outsmarted more than one hundred agents of the DEA."

They stayed three steps behind Guzmán
They looked for him in Los Cabos
But he was already in Culiacán!

The corrido got one thing right: Chapo was back on his home turf in the mountains. In the following months, the FBI continued to obtain Chapo's new numbers, then DEA Mexico would ping them to rural areas of Sinaloa, and later the nearby state of Nayarit. DEA Mexico then passed the intelligence to the Federal Police, which conducted additional raids, only to find that the target phone was not in the hands of Chapo at all. Instead the phone was being used by some low-level cartel employee who was only forwarding messages to Chapo's actual device.

And now no one had that number.

That was because Guzmán was employing the technique of a "mirror." It was the first time Diego and I had heard of Chapo using one. Mirroring wasn't a complex way of dodging law enforcement surveillance, but it was highly effective, if done correctly.

"Always one step ahead," I told Diego. "Chapo's smart—restructuring his communications as soon as he returned safely to Sinaloa."

After continued failed attempts in which they hit only the mirror (the low-level employee holding the target phone), the FBI's numbers began to dry up, and DEA Mexico, along with the Federal Police, decided to throw in the towel. DEA Mexico even closed the case file, and it didn't appear as if anyone was reopening a Chapo Guzmán investigation anytime soon.

BEFORE I EVEN PUT IN for the position, I knew I'd be ending a once-in-a-lifetime partnership. As much as Diego would've loved to investigate cartels south of the border, he wasn't a fed; he was a Task Force officer—a local Mesa, Arizona, detective—and couldn't reside in another country. The invitation to my going-away party had a picture of Diego and me together in our tactical vests just after we'd finished a big raid, smiles on our faces, the tangerine shades of the setting Arizona sun behind us.

Over the years, if our casework took us to Southern California, Diego and I would often zip down to Tijuana to take in even more of the Mexican culture I'd come to love. We'd listen to mariachi, banda, and norteño, then swing over to the strip clubs at 3 a.m., before grabbing a handful of street tacos and heading back across the border. For me, it was all part of learning the culture, deepening my understanding of a world I'd submersed myself in since that first night at Mariscos Navolato when I heard "El Niño de La Tuna" and began educating myself on the Mexican cartels.

I would *never* have gone to Tijuana without Diego. We weren't tourists, after all—a DEA agent and a detective from an elite counter-narcotics task force—and if anyone knew who we *really* were, especially with the heavyweight cartel drug and money-laundering cases we were working, we'd have made extremely vulnerable targets.

For the going-away party, several of my buddies from back home flew in: even my old sergeant from the sheriff's office. The celebration kicked off at one of San Diego's craft-beer bistros—a night of war stories, a running slide show of my time with Team 3, and the requisite plaques and framed photos—but the party didn't end when the bosses went home. Instead, at 2 a.m., I grabbed my closest friends and suggested we pop down to Mexico. But just as we were about to leave, Diego stared at his buzzing iPhone. "Fuck—family emergency," he said abruptly, hugging me. "Sorry, dude—gotta bounce."

My friends and I jammed into a cab and raced to the border. A taxi full of gringos, and no Diego as our guide. I had heard the cold click of the border pedestrian gates close behind me many times before, but now it was all on me: I would have to do all the talking and navigating.

Fresh out of language school, my Spanish was good enough—my teacher was from Guadalajara, so my accent was consistent with the locals'. But my vocabulary was still so limited that I often found myself getting knee-deep in conversations I just couldn't get out of until I'd end the interaction abruptly with a nod and a *"gracias."*

Somehow I managed to lead my Kansas buddies through the night, tossing back shots of Don Julio, rolling over to a streetside taco stand, mowing down al pastor on the spit, and walking back across the border into California just as the sun was cresting the mountains to the east. *Diego should've been here to see this*, I thought, but then I realized it was almost a rite of passage that I was able now to handle Tijuana on my own.

THE FOLLOWING DAY I was at the San Diego International Airport with my family, lugging the cart loaded up with our suitcases and carry-on bags through the terminal to check in. I was just another dad, hands full of passports, boarding passes—and my sons tugging at my elbow.

Whatever risks lay ahead, I was more certain than ever I'd made the right decision.

The plane ascended through the clouds—my sons fell fast asleep on my shoulders—and, for the next couple of hours, at least, hunting down Chapo Guzmán was the furthest thing from my mind.

DF

MY FAMILY AND I touched down in Mexico the last week of May 2012. The sprawling metropolis—with twenty-six million people, the largest city in the Western Hemisphere—was rarely referred to by locals as "Ciudad de México." To the natives it was El Distrito Federal ("DF") or, owing to the ever-present layer of smog, El Humo ("The Smoke").

At the embassy, I'd initially been assigned to the Money Laundering Group. The Sinaloa Cartel desk was run by a special agent who was burned out, especially after the Cabo fiasco. After a few months, I convinced management to transfer me over from Money Laundering to the Enforcement Group. The following morning, I sat down for breakfast with my new colleagues and group supervisor at Agave, a café known for its machaca con huevo and freshly baked pan dulce.

Before my arrival, the system had been inefficient. Most DEA special agents were working leads on multiple cartels: Sinaloa, the Zetas, the Gulf Cartel, the Beltrán-Leyvas, the Knights Templar . . . My group supervisor knew that this lack of focus was highly counterproductive. The Mexico City Country Office was such a hive of activity that no special agent could become a subject-matter expert on one particular cartel, because they were constantly working *all* of them.

So, in one of my first meetings with my new team, we began a reorganization. We went around the table to focus our assignments, and when it came to the Sinaloa Cartel, the assigned agent spoke up immediately, nodding at me.

"You can have this *desmadre* of a case," he said. "I'm *done*. The Mexicans couldn't catch Chapo if he was standing at the fuckin' Starbucks across the street from the embassy."

"Sure, I'll take it," I said, trying to contain my excitement.

"Adelante y suerte, amigo." Go ahead, and good luck.

At that moment, my mind drifted far from the meeting: I tried to picture Chapo eating breakfast, too, in a mountain hideaway or on some ranch in the heart of Sinaloa . . . *Somewhere* in Mexico. At least we were now on the same soil.

The task in front of me was daunting. After all the failed capture operations, all the years of near misses, I knew that Chapo must have learned from his mistakes. He had the resources, the money, and the street smarts to secrete himself so deeply into his underworld that it would now be extremely difficult—perhaps even impossible—to take him unaware.

He has eleven years of hard study on me, I thought as the meeting wrapped up. *I've got a hell of a lot of catching up to do.*

AS I SETTLED INTO my work at the embassy, I ran into Thomas McAllister, the DEA regional director for North and Central Americas Region (NCAR). He gave me a piercing look.

"Hogan, I'm told if anyone can catch Chapo, it's you . . ."

It was more a question than a statement, and I felt my face

flush. I knew where this had come from: my first group supervisor, back in Phoenix, had worked with McAllister at DEA headquarters and knew exactly how relentless and methodical I was when pursuing the targets of my investigations.

"We'll see, sir," I said, smiling. "I'll give it my best shot."

One thing I had vowed: I was not going to fall into the trap of believing all the legends and hype. Even some of my DEA colleagues had lost hope, so I disengaged emotionally from the Chapo mythology, instead focusing on it from the most basic policing perspective. No criminal was *impossible* to capture, after all, the failed Cabo operation proved that Chapo was more vulnerable now than ever.

NO SOONER HAD I settled into my seat in the Enforcement Group than I was assigned to be the DEA liaison on a case dominating all the Mexican headlines: a drug-related murder in broad daylight inside the terminal at the Mexico City International Airport. The airport was known to be among the most corrupt in the world; inbound flights from the Andes, especially Peru, almost always had cocaine hidden in the cargo. What made the incident all the more shocking was that the murders involved uniformed Mexican cops shooting fellow Mexican cops.

Two Federal Police officers assigned to the airport were just getting off their shift, walking through Terminal 2. They were attempting to smuggle several kilos of cocaine hidden underneath the navy blue jacket of one of the officers—marked POLICÍA FEDERAL in white on the back—when they were approached by three PF officers coming *on* shift, who had become suspicious of them.

A quick argument ensued between the two groups as they stood near the public food court. The dirty cops drew their service pistols and began gunning down the honest cops. One was executed with a point-blank shot to the head; two others were hit and died. To outsiders, the carnage looked like a terrorist attack; horrified travelers were screaming and scrambling for cover. Meanwhile, the corrupt officers took off running through the terminal, jumped in a truck, and sped away.

"Can you believe this shit?" I turned toward a senior agent in the group. "Blue-on-blue in broad daylight in the middle of an international airport? Who are these guys?"

The agent wasn't fazed; he didn't even look up from his computer screen.

"Bienvenidos," he said. Welcome to Mexico.

Though I tried to assist the Federal Police and the PGR—the Mexican Attorney General's Office—tracking down the murderers, I soon came face-to-face with a harsh reality: there were too many layers of corruption. The investigation into the blue-on-blue killings faltered and eventually went cold. It was a stark introduction for me—I saw firsthand, within weeks of my new assignment, why fewer than five percent of homicides in Mexico are ever solved.

OF ALL THE STUNNING cases of corruption and violence in Latin America, few lingered like the case of DEA agent Enrique "Kiki" Camarena, who vanished on a busy street in Guadalajara in 1985 while walking to meet his wife for lunch. Camarena's body wasn't found for nearly a month. When it was, it was discovered that his skull, jaw, nose, cheekbones, and windpipe had all been crushed; his ribs had been broken and he'd been viciously tortured; he was even sodomized with a broom handle. Perhaps worst of all, his head had been drilled with a screwdriver, and he'd been buried in a shallow grave while still breathing.

Kiki Camarena's disappearance became a major international incident and heavily strained relations between the United States and Mexico—the US government offered a $5 million reward for the arrest of the murderers.

When I arrived at DEA's Mexico City Country Office more than twenty-five years later, the circumstances surrounding Camarena's death had not been forgotten. His memory was kept vividly alive. Along the main embassy hallway, a conference room dedicated to the slain agent—we referred to it simply as the Kiki Room—featured a small bust of Camarena and a plaque. Convicted of Kiki's torture-murder was none other than Miguel Ángel Félix Gallardo, "El Padrino," a former Federal Police officer turned godfather of the Guadalajara Cartel—and Joaquín Guzmán's mentor in the narcotics business.*

LOOKING BEYOND THE DEEP-ROOTED history of violence in Mexico, I tried to give my wife and young sons the best life possible in the capital under the enormously high-stress circumstances. The DEA assigned us a spacious three-bedroom apartment in La Condesa, the city center's hippest neighborhood—I suppose you could compare it to Paris's Latin Quarter or Manhattan's SoHo—home to young businesspeople, artists, and students. It was also close to the US embassy, on Paseo de la Reforma, so just a fifteen-minute drive to work.

---

* Miguel Ángel Félix Gallardo and two other Guadalajara Cartel kingpins, Ernesto Fonseca Carrillo and Rafael Caro Quintero, were all ultimately convicted in connection with the Kiki Camarena murder.

We loved the neighborhood, full of tree-lined streets shading the 1920s architecture: restaurants, cafés, boutiques, galleries, and lively open-air markets on Sundays.

But it was difficult for me to enjoy the vibrant life of the city: my head was constantly on a swivel. *Street-cop mode.* It was second nature to be watching my back—I'd done so since I was twenty-one and on patrol with the sheriff's office—but in Mexico, there never seemed to be a moment's rest. I was always checking for tails and surveillance by members of the cartel, street thugs, or even the Mexican government. When I left our apartment at 7 a.m., walking out to my Chevy Tahoe, I'd study all the other vehicles on the street. Which cars were new to the block? Which ones seemed out of place? Which cars had someone sitting inside them? I'd even memorize makes, models, and plate numbers.

Whenever we went to a new neighborhood, my wife knew that there was no point talking to me. I was too busy scanning the streets, looking hard at the faces of pedestrians, taxi drivers, deliverymen—anyone, in fact, within shooting distance.

After just a few weeks in DF, my wife had also learned the techniques of constant risk assessment: look everyone on the sidewalk in the eyes quickly to judge them, and decide: threat or not? She and our young sons were always on the street, at the park, shopping, or meeting friends. There was crime all over DF, but of a random nature: we'd hear reports of an embassy employee being held up at gunpoint for his gold watch in a local restaurant in our neighborhood, or a lady out pushing her kid in a stroller having her handbag snatched.

But there were plenty of great things about living in Mexico, too. We especially loved the city's street food: tacos de canasta, tlacoyos, elote (sweet corn in a cup with melting butter topped with a dab of mayonnaise and chili powder). But best of all were the camotes—sweet potatoes—from a vendor who'd come around every week at sundown, pushing his old squeaky metal cart.

The guy looked as if he'd been working in the sun all day, face golden brown, covered in beads of perspiration from pushing his wood-burning stove up and down the streets. The pressure of the smoke and heat from the fire would sound a steam whistle, like an old locomotive in a Western movie. You could hear the sound coming from blocks away, even if you were indoors. One of my sons would shout:

"Daddy, the camote man!"

We'd throw on our shoes and run outside. Sometimes the camote man would be gone, vanishing in the shadows down side streets before he'd sound his whistle again, directing us where to run. Once we hunted him down, he'd pull out a drawer full of large sweet potatoes roasted over the wood fire and let my sons pick out the best-looking ones, then he'd slice them lengthwise and drizzle condensed milk over the top and add a heavy sprinkle of cinnamon and sugar—a bargain at just twenty-five pesos.

Even in those sweet moments, as much as I tried to mask it from my sons, I was on edge. Children were always the most vulnerable for kidnapping—we even had one neighbor, a "self-made millionaire," who'd fly his daughter to school in a private helicopter every weekday.

It wasn't strange to see the latest Ferraris and Porsches ripping through our neighborhood streets—though anything lavish and excessive in the capital reeked of a narco connection. There was an estimated $40 billion a year in drug money flowing through the country's economy, and it had to trickle down somewhere.

I was constantly reminded of a remark that I'd heard from a local journalist in DF: "Everything is fine in Mexico until *suddenly* it's not." The expression captured it all in chilling simplicity. "You're living your life happily and then one day you're dead."

CHAPO HAD FINALLY BECOME a household name in the United States, designated Public Enemy Number One by the Chicago Crime Commission—the first outlaw to earn that title since Al Capone. And while I was glad this label drew more attention to Guzmán's name and his criminal activity, it did little, from an investigative standpoint, to assist with a capture.

At my embassy desk, I spent day after day sorting intel on Guzmán, dissecting every old file I could get my hands on. The freshest leads were the ones that came from the notebook pages, ledgers, business cards, and even the pocket trash left behind at the mansion after the raid in Cabo San Lucas. It was grueling analysis—the sort of work despised by most DEA agents—but I'd find even the slightest variation of a nickname or the subscriber to a phone invaluable, and when I found something, it hit me like a shot of adrenaline.

Exploit. Exploit. Exploit.

My life soon became an endless blur of digits. I had become obsessed with numbers. I was constantly memorizing any phone number, any BlackBerry or PIN number I could find. I couldn't remember my grandma's birthday, but I had Chapo's pilot's phone number on the tip of my tongue. The other agents in the group would ask why I was always consumed with analyzing phone numbers and PINs.

Numbers, unlike people, never lie.

———

NOT ONLY DID CHAPO and Picudo leave crumbs behind in Cabo San Lucas, but they took off so quickly that Chapo never had time to grab his tactical go-bag containing his forest-green armored vest, black AR-15 rifle equipped with a grenade launcher, and six hand grenades.

Diego and I confirmed that Guzmán had even cut himself on a fence, drawing blood, but was now resting comfortably back across the Sea of Cortez in Sinaloa. For Chapo, this was as close as he had ever been to capture since his breakout from the Puente Grande penitentiary. I knew he was becoming complacent if he felt he could spend time in such a popular resort city, especially one swarming with foreign tourists. And clearly he wasn't escorted by hundreds of bodyguards driving fleets of black armored SUVs with tinted windows, as people had claimed. It was intel that was still widely believed, including by the US intelligence community in Mexico.

Once in a while I'd share my findings with the Mexican Federal Police team that had worked the leads after the February raid in Cabo, and PF would give me any bits and pieces of intelligence they had collected. I'd end up divulging far more information than I'd receive, but I reasoned that *some* Mexican intel was better than none.

Then it was back to digging through the active phone numbers of Chapo's pilots, family, girlfriends—often never raising my head above my computer screen until another agent would make a sarcastic comment.

"Why you wasting your time, Hogan? What's the endgame? The Mexicans will never catch Chapo."

Even my bosses were skeptical as they eyed the massive charts I'd pinned on the wall, linking multi-ton cocaine seizures in Ecuador directly to Chapo's lieutenants.

"Cuando? Cuando?" my boss would often yell as he walked by my desk, demanding to know when—if ever—I was going to show something for all the effort.

"Paciencia, jefe, paciencia," I would say. "Have some patience, boss."

EVERY NIGHT I LEFT the embassy, my head was back on a swivel. DF was a constant swarm of cars and pedestrians, and I knew that at any hour of the day or night someone could be watching me.

Or, worse, trying to follow me.

I was headed home one evening at dusk, driving away from the embassy on side streets in my Tahoe. As I took my first right, I made a mental note of the vehicles behind me that did the same.

Blue Chevy Malibu. White Nissan Sentra.

I took a left at the next light; the white Sentra did the same. In my rearview mirror I could make out the sharp cheekbones, dark eyes, and thick brow of the driver.

Was it the same guy with the blade scar on his cheek on that hot afternoon money drop in Plaza Satélite? It sure looked like him . . .

I couldn't be sure, but I hit the gas hard—another left and then a quick right, making sure I cleaned my tail of the Nissan.

But I felt relatively safe in my Chevy Tahoe, with its two-inch-thick bulletproof glass. It was so heavy from all that level 3 armor that just a gentle tap on the gas pedal made it sound like it was going eighty-five miles an hour. A seasoned DEA agent at the embassy would say, in his heavy West Texan accent, "Them babies run like scalded apes." Surveillance by Chapo's people would be nearly impossible, the way I drove the thing—after just a month in-country I knew all of the shortcuts home, and regularly changed my route to and from work.

IT WAS A BLAZING afternoon in August 2012, and Tom Greene, an agent in my group—working the Beltrán-Leyva DTO—was agitated, constantly checking his BlackBerry.

"Funny, he's not responding," Tom told me. Greene had just returned from meeting with his informant, El Potrillo ("The Colt"), a twenty-six-year-old with a heavyset frame and a long, thin face, from just outside Mexico City. Tom and Potrillo had met a few minutes earlier at a small café-bookstore called El Tiempo, just a block from the embassy in the Zona Rosa neighborhood.

"I've sent him a shitload of messages," Greene said. "Kid always texts right back."

It didn't seem like a big deal, so Tom and I went to lunch in the embassy cafeteria. As we were standing in line with our trays, we overheard one of the cashiers speaking Spanish: "Did you hear? Horrible. There was just a shooting over in Zona Rosa . . ."

Just west of the historic center of Mexico City, Zona Rosa was a perfect place to meet a confidential informant, because it was one of the capital's most bustling and vibrant neighborhoods—full of nightclubs, after-hours joints, and gay bars. After meeting with El Potrillo, Greene had seen a couple of suspicious guys on the street, one in a car and another walking slowly down the sidewalk, but he had thought little of it. His informant followed protocol, waiting to exit El Tiempo until Greene was long gone.

El Potrillo had only taken a few steps down the busy sidewalk when a motorcycle pulled up alongside him. There were two male riders in full-faced black helmets. The rear rider got off the back of the Yamaha, walked calmly up behind El Potrillo, and shot him in the back of the head six times. Five of the bullets had been super-fluous; El Potrillo was most likely brain-dead by the time he hit the pavement. The assassin jumped on the back of the motorcycle and sped off. The killers had used a classic sicario technique, imported to the Mexican capital by Colombian hit squads.

I walked by the spot a couple of days later and could still see the bloodstains—now the color of dried wine—on the sidewalk.

The police investigation went nowhere; none of the witnesses would cooperate. The assassins' Yamaha had had no license plates. In fact, not one piece of evidence was noted by the local cops besides the time and place of the shooting. It quickly became another stat: one of the tens of thousands of drug-related homicides that remained unscrutinized and unsolved.

AFTER GREENE DEALT WITH a few days of trauma in the DEA office, I found that life, strangely enough, went back to normal. The execution of El Potrillo was just another nightmare moment that Mexico City belched out daily, like the clouds of smog that hover over the metropolis—and yet another constant reminder that I could be shot point-blank in the back of the head at any moment, too, if I didn't remain hypervigilant.

If anything leaked out to the wrong people (narcos, dirty cops, even some greedy civilian looking for a payday), if anyone were to learn who I was actually targeting or the work I'd been doing for more than six years, it wouldn't be some informant bleeding out in the streets of Zona Rosa—I would be another Kiki Camarena.

Several weeks later, two CIA employees were driving to a military installation on the outskirts of the city in a Chevy Tahoe with diplomatic license plates—an armored vehicle identical to mine—when they were ambushed by two vehicles loaded with gunmen. The Tahoe was sprayed with more than a hundred machine gun rounds. The bad guys—it turned out they were rogue Mexican Federal Police—laid fire in such rapid succession that the bullets pierced the armor, striking the two CIA employees inside. But unlike Special Agent Zapata, they survived—they kept the Tahoe crawling on the metal rims until it could go no further.

I studied the photographs: that Tahoe looked like it had just driven out of a firefight in Fallujah.

I walked out of the office that very night and opened the door to my own Tahoe, my left eye twitching, and felt a bristling cold shiver, in spite of the midsummer heat, knowing that I—or any other DEA agent in the embassy—could be the next target of a murder.

BADGELESS

HE WASN'T DIEGO. But then again, who was? Homeland Security Investigations Special Agent Brady Fallon brought his own unique skills to the table, and my partnership with him was almost as unlikely as the one I'd formed with Diego. Not ethnically—we were both Irish Americans; Brady had been born in Baltimore, studied finance in college abroad, and become a fed immediately after the September 11 terrorist attacks. What made our connection so unique was that agents from the Drug Enforcement Administration and HSI typically *detested* each other.

At the executive level, in Washington, DC, communication between the agencies was done through snail mail—agents wouldn't even pick up the phone to talk to each other. There had been deep-seated enmity even before the Office of Homeland Security, established after 9/11, became the cabinet-level Department of Homeland Security on November 25, 2002. It was much like the dysfunction between the FBI and the CIA—competition and a desire for personal credit overrode cooperation and common sense.

Special agents from DEA and HSI would typically get into territorial pissing matches. . . . Then a headline-hungry assistant US attorney might get added to the mix with a case like this—the potential to arrest and charge the world's most wanted narcotrafficker—and the investigation would come crashing down in a matter of weeks. That was precisely why no one had gotten a bead on Chapo in the twelve years since he'd broken out of prison.

My relationship with Brady began in April 2013, with what I thought would be a routine deconfliction hit, just another DEA office or US federal agency investigating the same BlackBerry PINs as I was. I called the agent whose name popped up on my screen: Brady Fallon, HSI—El Paso Field Office.

"So tell me, does your guy '06' also go by 'Sixto'? And your 'El 81'—has anyone ever called him 'Araña'?" I asked.

I could imagine what Brady was thinking: *Great—another DEA cowboy who wants to come in and sweep up our entire case . . .*

I could hear a muffled voice as Brady yelled to a few of his Homeland Security guys in the background, then he came back on the phone and said, "Yes, we have them referred to by those names. Araña comes up—so does Sixto. Why?"

"Listen," I said, "I don't know if you guys realize it, but you're sitting on two of Chapo Guzmán's most trusted pilots."

Not only were Brady and I both targeting the same PINs—for Sixto and Araña—but there was another PIN I'd found while piecing together go-fast boats smuggling tons of cocaine off the coast of Ecuador, bound for Mexico's west coast. Brady had the user name of that PIN listed as "Ofis-5," and said that whoever was on the other end of that device was placing some serious orders to traffickers in Guatemala, Colombia, and Ecuador. And the recipients would always acknowledge the message with the words *"Saludos a generente."* Greetings to the manager.

"Sometimes these guys will address their messages to 'El Señor,' too," said Brady.

"Yeah, El Señor," I repeated.

That level of respect almost surely meant it was a reference to Chapo.

Working along the Texas border across from Ciudad Juárez—a narco war zone, and the city with the highest murder rate in the world—Brady had had his share of unpleasant dealings with DEA agents.

At one point, he had invited a DEA agent from the El Paso, Texas, office to help with their case; he was told that the only way DEA was going to help was if they could lead the investigation; Brady wasn't going to let that happen and slammed the door. So he was still skeptical.

"How do I know you're not just going to run off with all of my intel?" Brady said.

I understood his concern. "You don't know me yet, but I pride myself on knowing everything there is to know about my targets and sharing it with guys who want to jump on board and work together."

I had closely studied the systemic failure of intel sharing between the FBI and the CIA—the catastrophic interagency dysfunction in the period before 9/11—and promised myself I'd never withhold information from another federal agency if it would further the investigation. I had learned early on—going back to Task Force days with Diego—that it was the tight relationships I fostered throughout my career that had helped me be successful in every case I'd led. No one could come along and tell me that they knew more than I did—because, frankly, they had never dug deep enough. This wasn't arrogance; it was just my thorough method of investigating.

After setting the tone with Brady, I immediately began to fill him in on everything I knew about the two pilots, Sixto and

Araña, and how Ofis-5 was connected to seizures down south that were directly linked to Chapo.

"This shit could be a gold mine," Brady said.

BRADY AND I WERE soon on the phone twice a week, comparing notes, phone numbers, and intercepts from the Ofis-5 wire.

We laughed about the far-fetched stories that filled the various government agency intel files. Chapo never had plastic surgery to disguise himself; he wasn't hiding in Buenos Aires; he wasn't living a life of luxury in the Venezuelan jungle, drinking tea and talking politics with Hugo Chávez. No one in the US government's alphabet soup—DEA, HSI, FBI, ATF, or CIA—had bothered to sift through all the stories to determine fact from fable. There was no coordination in targeting efforts, and everyone slowly began to believe all the myths, repeated and retyped often enough that they were regarded as gospel.

In time, I brought Brady up to speed on the success Diego and I had had back in Phoenix, how far we had come in our Team America investigation, and the failed operation in Cabo. I sent him a photo of Chapo, flanked by three women and appearing to be in good health.

Diego had sent me the photo after it was found on an abandoned BlackBerry taken from the Cabo mansion. It was the most up-to-date photograph that the US and Mexican governments had of the top drug fugitive in the world, and it had never been seen by the public.

From inside the embassy, I began to GPS-ping Ofis-5.

Boom.

Within seconds I had pinpointed the device in Durango, east of Sinaloa. I couldn't believe that Guzmán would be hanging out in the middle of a busy city again—but who knew? With Chapo, anything was possible.

I explained to Brady the way Chapo had restructured his communications network after returning safely to Sinaloa from Baja.

"That Ofis-5 device is most likely a mirror," I said. "Short for *oficina*."

"*Oficinas*," Brady said. "Makes sense. They're functioning like offices."

"Exactly—Chapo's often referred to his mirror locations as offices."

They were grand-sounding, but Brady and I later learned that they were really just cinder-block apartments—shitholes—with the "office" worker thumb-typing thousands of messages, never seeing daylight, surviving on a pot of refried beans and an occasional Burger King Whopper. For sixteen hours a day, the mirror would relay all communications and send them to the intended recipients: they acted like a central switchboard for the cartel and also served to limit Chapo's direct communication with anyone.

"So you think he's still insulating himself?" Brady asked.

"Yes, he's insulating himself well. Now it's just a matter of how many layers are between him and us. For now, these offices are our key."

It reminded me of the old-school American Mafia walk-and-talk—always insulate the boss from direct communication.

And then I went back to my work, opening a fresh Google Map of Mexico on my MacBook and placing my first red-pin marker on the coordinates of Ofis-5 in Durango.

———

ON THE EVENING OF April 4, Brady and I learned through a DEA agent in New York—working a confidential source—that Chapo would be celebrating his fifty-sixth birthday, surrounded by friends and family, at a ranch in his birthplace of La Tuna, the hacienda perched high in the Sierra Madre in the state of Sinaloa. *Feliz cumple!* Birthday messages streamed in for El Señor. It was the first time we knew where Chapo was located since he'd fled Cabo.

But we couldn't act on the intel.

"It's too early—and far too risky at this point—to organize a capture operation, and I don't even know who we could trust in the Mexican counterparts," I said.

The same scenario had been tried, and had failed, numerous times. For years, DEA agents working with their Mexican law enforcement counterparts would act on viable intelligence derived from confidential sources reporting where Chapo would be. Sometimes it was a large fiesta in the mountains, other times a small meeting behind closed doors on some trusted lieutenant's ranch.

I had studied the history: the capture ops were always rushed and reactive. The DEA agents would typically have one or two days' advance notice, grab the first Mexican counterpart willing to risk his unit, and mount a fast capture op. Invariably Chapo would catch wind of the plan a day or hours beforehand and vanish.

No one knew where the leaks and tip-offs came from, but Chapo always had plenty of advance notice. Each time the Mexicans came up empty-handed, the DEA would blame systemic corruption, tuck their tails between their legs, and return home. There were never any persistent and sustained operations to follow, because no one had put in the time and done the work to see the big picture—to know where Chapo had come from, let alone where he was headed next.

Up to this point, it had all been a crapshoot—haphazard and improvised—with each miss bolstering Chapo's reputation as untouchable.

"So other than you," Brady said, "who's targeting him?"

I knew the answer to the question, but I let the silence on the line linger for effect.

"No one."

"You've got to be shitting me?"

"Not kidding," I laughed. "Just me."

"Unbelievable," Brady said.

"There's a lot of DEA offices targeting the upper echelon of the Sinaloa Cartel. They're all trying to work an angle to penetrate Chapo's inner circle. I'm working with all of them. Sure, they each have a piece of the puzzle, but they're not close enough yet. You and me, we're it, man. If anyone has a chance to catch Chapo, it's us."

Brady and I knew that the potential breakthroughs would be buried in those line sheets, in those messages.

"We just gotta keep digging," I said.

"Easy enough to target one trafficker," Brady said. "But that sounds like it's not gonna work with Chaps."

"No," I said. "Never has. We need to exploit his entire inner circle. The lieutenants, enforcers, couriers, pilots, lawyers, and accountants. His sons, nephews, cousins, wives, girlfriends. Even his cook and maids."

The key investigative strategy would be to simultaneously

exploit *multiple* avenues leading to El Señor. If one line failed, we'd have several backup options, allowing us to pivot without losing any momentum.

"Chapo's been one step ahead ever since he broke out of Puente Grande," I said. "He understands who's hunting him. He understands *how* we operate. This guy's no fool. He's always watching his back."

For years Chapo had outsmarted some of the best agents on both sides of the border, but in Brady Fallon I felt I had a fellow federal agent who shared my conviction and determination. Together I hoped that we might have the right stuff to outsmart Chapo at his own game.

IN LESS THAN a month, Brady's HSI team in El Paso was intercepting two more *oficina* BlackBerry devices that we'd identified.

Brady and I determined that each office was in communication with anywhere from five to ten members of Chapo's inner circle, and each office was responsible for sending Chapo's orders to its designated contacts.

On a whiteboard in the embassy, I drew a line from each office tasked with communicating orders from above to the actual lead cartel operators—Chapo's core workforce, and the men representing his authority in Mexico and Central and South American countries:

Office-1—Tocallo
Office-3—Lic-F, Lic Oro
Office-5—Chuy, Pepe, Fresa, Turbo

For the first time ever, Brady and I were peering through a keyhole into Chapo's inner world, witnessing the volume of activity flowing somehow—through all the mirrors—from the office devices back up to Chapo.

Brady and his team had already done a ton of legwork over the course of several months and were well versed on several of the key players:

Chuy was an operator based in Guatemala who coordinated large cocaine loads coming up from Colombia and Venezuela. He would work with Chapo's pilots, like Sixto, to get the loads across the southern border into Mexico.

Pepe worked right at the source of the coke pipeline—deep in the jungles of Colombia—trying to secure thousands of kilos of cocaine base, which he would then send north in go-fast boats staged on the coast near the Colombia–Ecuador border. In the line sheets, it was evident that Pepe was a hard worker and reliable; he always provided updates on his progress to El Señor, mirrored through Office-5.

Fresa was the chief operator based in Ecuador who had the responsibility of finding clandestine airstrips in rural areas so he could receive loads of cocaine base, to be flown out of the country in private planes. Brady and I could see that Fresa was nowhere near as reliable as Pepe.

"This guy Fresa down in Ecuador is constantly bitching about not getting paid," said Brady.

"Yeah, I see that. And *el generente* isn't real impressed with his work. Fresa better get on point."

Pissing off the manager of the world's largest drug cartel usually didn't end with just a verbal reprimand.

TOCALLO? LATE INTO THE NIGHTS, I kept drilling down into the line sheets on Office-1. But that word kept leaping out at me from the blur of the daily back-and-forth among all the office devices.

I'd first heard the word *tocayo*—"namesake"—years ago from Diego during our Task Force years in Phoenix. In many Mexican families, Diego had told me, Tocayo—often misspelled Tocallo—was an affectionate way of referring to someone who has the same name as you.

"Tocallo on Office-1," I told Brady. "I'd bet anything this Tocallo is gonna turn out to be Iván."

"Iván?" Brady said.

"Yeah, Iván Archivaldo Guzmán Salazar."

"His son . . . I remember."

"You can tell just by the way that they're speaking. There's a level of respect they have for each other. And it's the first numbered office device—why wouldn't Chapo designate the first office to his number-one son?"

"What've you got on him?"

"Iván's known to be moving ton quantities of weed from Sinaloa up through Sonora and then into Tijuana and Nogales. Chapo and Iván share the same middle name, Archivaldo. That's a namesake. *Tocallo*. Can't be a coincidence," I said. "It's gotta be Iván."

Iván was one of Chapo's sons by his first wife, María Alejandrina Salazar Hernández. Born in 1983, and often referred to as "Chapito" after Edgar was murdered, Iván eventually took over as Chapo's heir. Now he was the most trusted son. Iván and his younger brother, Jesús Alfredo Guzmán Salazar, spent most of their earlier years bouncing between Culiacán and Guadalajara, living the life of the ultimate narco juniors: throwing lavish parties and driving rare European sports cars. Now Iván and Alfredo were running their own semiautonomous DTO and helping their father out with whatever he needed. Alfredo and Chapo were federally indicted together in Chicago in 2009 on drug-trafficking and money-laundering conspiracy charges.

While their father was trying to keep a low profile, Iván and Alfredo couldn't get enough of the narco life, buying the world's most exclusive cars—Lamborghinis, Ferraris, Shelby Mustangs, even a rare silver Mercedes-Benz SLR McLaren, with batwing doors, which was imported from England and could go from zero to sixty in 3.4 seconds—all ordered in the United States and Europe through straw purchasers. They also bought private aircraft, though they never paid full price—just like their father, the sons always haggled to get the best deal. They wore oversize Swiss watches, carried bejeweled pistols, and even had wild A1 Savannah cats, imported from Africa, living as their pets in Culiacán.

THE MESSAGES KEPT POURING into HSI's El Paso office.

"These guys are non-fucking-stop," Brady said. "My translators can hardly keep up with the line sheets."

Every day in Mexico, I would receive a fresh batch of the latest lines and spend my entire day reading and rereading them, trying to decipher just a sliver of Chapo's global drug communications. With all of the fast-paced activity of cocaine loads bouncing north from country to country, it was easy to get distracted, but now that I had access to the "tolls" (call detail records) from multiple office devices, I could do what I'd learned to do best.

We've got to move up the ladder, I would tell myself. *Exploit . . .*

If the office devices were all mirrors, then who was above them?

I ran a quick frequency report on the offices, which provided the most common PIN in contact with each of them, and quickly noticed a common thread. I knew from all of my years analyzing numbers that the PIN in most frequent contact with the target would likely lead to the boss. With normal phone users it would typically be a spouse, significant other, or parent, but in the drug trade the most frequent contact PIN was invariably the shot caller, the boss. One hundred, two hundred, or even three hundred messages were sent daily to that most frequent PIN.

26B8473D

I took that most frequently contacted PIN and pinged it.

If this was indeed Chapo's personal PIN, I'd have a pretty good idea based on location. The results came back within seconds.

Right in the heart of Ciudad de Durango, the capital of Durango State, again.

"Shit," I told Brady. "Looks like just another mirror."

The username of the PIN was "Telcel." Brady and I dubbed this next layer "Second-Tier."

"You guys are writing for Second-Tier, right?" I asked Brady.

"Yeah, one step ahead of you," Brady said. "I've already got my guys on it. Have you read what's been coming in on Office-3 lately?"

"Just diving in this morning," I said.

In the line sheets I saw that Office-3 was in regular contact with all the "Lics" in the organization. Lic was shorthand for *licenciado*.

I knew from my casework in Phoenix with Diego that

licenciado—literally "licensed title"—could refer to anyone who'd earned an advanced degree: engineers, psychologists, architects. But in common Mexican usage, a *licenciado* was generally a lawyer or someone with any higher education. No one could be sure, but one of the principal advisers in the line sheets went by the name Lic Oro.

Filtering the message through Office-3 and then down to Lic Oro, El Señor would ask for the status of various court cases involving significant Sinaloa Cartel heavyweights who'd recently been arrested.

One of the most pressing legal cases involved a player Lic Oro referred to as "El Suegro."

I kept poring over those line sheets, seeing references to the case involving El Suegro (Spanish for "father-in-law").

Driving home from the embassy that night, I remembered one family barbecue in the suburbs of Phoenix where Diego had introduced me to his wife's father—using that title, *"mi suegro."* And during the Team America operation, learning that Carlos Torres-Ramos and Chapo were *consuegros* . . .

And then I suddenly understood that this El Suegro could be only one man: Inés Coronel Barreras, the father of Emma Coronel Aispuro, Chapo's young beauty-queen wife.

Emma was actually an American citizen—she was born in California in 1989—and had grown up in a remote Durango village called La Angostura. She'd married Guzmán when she was just eighteen, when Chapo was over fifty. Brady and I knew her history well—she'd caught Chapo's eye after winning some local pageant at La Gran Feria del Café y la Guayaba; her father was a cattle rancher and heavy hitter within the Sinaloa Cartel. In fact, on April 30, Inés Coronel Barreras had been arrested by the Mexican

Federal Police along the US border—in Agua Prieta, Sonora—for running a marijuana-and-cocaine-trafficking distribution cell responsible for smuggling large loads into Arizona.

Tocallo: namesake.

Had to be Iván.

El Suegro: father-in-law.

Had to be Inés.

As careful as the communications insulation system was, the nicknames and aliases were clear tells.

The names left little doubt: Brady and I were almost certain who El Señor was, the man at the top of this BlackBerry chain. The boss putting out the orders through the mirror devices—however many layers of them there were, and we still weren't sure—had to be Chapo Guzmán himself.

Once I got to my apartment in Condesa that evening, I poured myself a double shot of bourbon on the rocks, slumped back in my living room armchair, and pulled my BlackBerry out of my pocket to write a message to Brady.

Iván—Tocallo.

Inés—El Suegro.

The net was tightening: a string of Spanish names was drawing us closer to El Señor.

I knew that standard texting wasn't secure over Mexican cell towers, so I typed Brady a message on WhatsApp.

"We've got momentum now," I wrote. "Tocallo. El Suegro. We're rolling. But we need to meet up. How soon can you come down here?"

———

TELECONFERENCING AND TEXTING for three months could accomplish only so much.

We had to meet face-to-face.

Less than a week later, I met Brady at the Mexico City airport, right there in Terminal 2, not more than fifty feet from the food court where the blue-on-blue Federal Police murders had taken place.

I recognized him immediately—he was five-ten, had a shaved head, and was wearing a dark gray jacket and Ray-Bans perched on his forehead. He was walking toward me with a pissed-off scowl, although, as it turned out, he wasn't in the least bit angry. We held a long stare, looking at each other—not as special agents from rival US federal government agencies, but as *men* with a shared vision for our investigation.

"Badgeless," I said.

Brady nodded.

"Badgeless."

We sealed the deal with a handshake, pulling each other chest-to-chest in a powerful hug.

TOP-TIER

BY THE SUMMER OF 2013 I was running point in Mexico for all DEA offices targeting the Sinaloa Cartel, coordinating with other DEA agents and federal prosecutors in San Diego, Los Angeles, Chicago, New York, and Miami.

I now had a much better grasp of the overall umbrella-like structure of the cartel. Chapo may have been *el jefe de jefes*—boss of bosses—but there were other DTOs nearly as large as Chapo's personal organization that operated under the wing of the Sinaloa Cartel. Next to Chapo, Ismael Zambada García, a.k.a. "El Mayo," Chapo's longtime senior partner, was the most prominent.

Any trafficker below Chapo and Mayo needed their direct blessing to work and share resources within their territory.

I was regularly on the phone with DEA agents stationed in Canada, Guatemala, Costa Rica, Panama, Colombia, and Ecuador, passing leads and sharing intel on the movements of Chapo's countless drug shipments. With all the moving parts, I realized I needed to call a coordination meeting of all the far-flung agents who saw only a sliver of the intelligence and investigative leads.

In August 2013, I brought all the agents together in Mexico City; there were too many to fit in the Kiki Room, so we all assembled in the embassy auditorium. Each office presented its case on PowerPoint; as they did so, I would periodically interject, highlighting the connections I'd made to other cases, giving everyone in the room a greater sense of the scale of the investigation.

"If *we're* not targeting Chapo," I said, wrapping up the meeting, "then who the hell is? We're it—the elite USG team. Right here in this room. The only thing lacking is self-belief. Chapo is no super-criminal. He's a man standing in the dirt somewhere in this very country. He's breathing the same air as us. Like any other drug kingpin, Chapo's vulnerable. He can be caught. But we all need to do our homework."

After the meeting, I was eager to sit down and meet a couple members of Brady's core team who'd accompanied him to Mexico and had been working rigorously behind the scenes.

Special Agent Joe Dawson was a heavyset, mid-thirties guy with straight brown hair long enough to tie in a ponytail, and he was wearing a gray button-down shirt and loose purple tie that made him look like a young tech exec from Silicon Valley. Joe, working closely with Brady, had taken the brunt of writing federal affidavits for all the office devices and the cartel operators we considered worthy of intercepting. Whenever I'd call El Paso, Joe would be working into the small hours of the night, sitting alone at his cubicle under one small desk lamp, jamming to Metallica, and typing and deciphering line sheets with me over the phone. Joe had a near-photographic memory and could instantly recall Chapo's activities after reading them just once.

At our meeting he said, "You see this guy called Vago in the line sheets on Office-5?"

"Vago? I saw that."

"Looks like he's getting ready to go on a rampage. Know who he is?"

"It's another alias for Cholo Iván."

I had already run "Cholo Iván" through our DEA databases: his real name was Orso Iván Gastélum Cruz. Chapo's top sicario and plaza boss in the northern Sinaloan city of Los Mochis, Cholo Iván was a scary trafficker even by the standards of Mexican drug cartels.

"And you saw him talking about a guy called Picudo?" I asked.

Joe nodded.

After Carlos Adrián Guardado Salcido, a.k.a. "El 50," died in a shoot-out with a local unit of the Mexican Army in August 2013, Picudo had stepped up to become Chapo's chief enforcer and plaza boss of Culiacán.

"You have Picudo's real name?" Joe asked.

I shook my head. "Picudo—it's Spanish for 'sharp' or 'thorny.' In Mexico it's a nickname for a badass, for a guy who's always looking for a fight. Picudo also goes by 'El 70.' Still working on getting his real name."

Picudo and Cholo Iván: these were two of the killers who gave Chapo his hold over the people of Sinaloa and through these sicarios, Guzmán could reign with violence.

In the days before my meetings with my DEA colleagues, Cholo Iván had been talking about killing "Los Cochinos"—a group from a rival cartel—in retaliation for the murder of Picudo's brother-in-law. Cholo Iván said they needed to attack Los Cochinos immediately, because elements in the Mexican government were aligning themselves with the rival cartel. Through the Office-5 mirror, Cholo Iván asked Miapa—slang for "my dad," a code name for Chapo—to send some more heavy artillery to him in Los Mochis.

We feared that bloodshed was coming.

———

OUR KEYHOLE INTO CHAPO'S world was rapidly expanding, but every few weeks—usually on a thirty-day schedule—*los pobrecitos* (the "poor ones"), as Brady and I called them, who ran all the office mirrors in Durango, would receive a bagful of new BlackBerrys, dropping all the old devices and instantly creating a logistical headache for us and our entire team.

Before we could intercept again, we had to try to identify the new office devices; then Joe would begin writing his affidavits. It was an arduous process that would take weeks to finish by the time an assistant US attorney reviewed the affidavit and Joe or Brady could get down to the federal courthouse in El Paso to have them signed by a US magistrate. Add another couple of days before HSI's tech group would "flip the switch." And all this had to be done for fifteen or twenty unique devices.

I realized it would take a small miracle for us to stay up long enough to break through all the layers of BlackBerry mirrors in Chapo's communications structure, let alone have enough time to be able to unravel Chapo's day-to-day operations.

But Brady and his handpicked team were not about to give up. We both knew that this entire investigation rested solely on his shoulders in El Paso and mine in Mexico City.

Fortunately, besides the staunch assistant US attorney they had working with them in El Paso, Brady had lined up another attorney with *juice*: a deputy chief for the US Attorney's Office in Washington, DC, Camila Defusio. A veteran prosecutor in her mid-forties, Defusio wasn't afraid of taking massive cartel cases, as long as they kept producing fruitful results. The Chapo case was right in her wheelhouse. She knew what needed to be done and would streamline HSI's affidavits even if it meant writing some herself. Like us, Camila saw the big picture, and Brady kept her well informed of our progress.

The intercepts on the Second-Tier device proved to be our vital portal.

Second-Tier may have been yet another mirror, but as soon as the BlackBerry went live, it was like a row of streetlamps turned on to illuminate a previously dark street. The crucial information in the line sheets grew exponentially. Not only were Offices 1, 3, and 5 routing their communications up to Second-Tier, but another three offices—numbers 2, 4, and 6—were doing the exact same thing.

THE INCOMING MESSAGES in the line sheets became an endless and intoxicating river. Brady was forwarding them to me by the pile— there were thousands of them. I could go six hours without moving or even getting up to take a leak. Each sentence from the offices exposed clues leading deeper into Chapo's secretive lair. I found I could be most productive when the other agents had left the embassy, from 6 p.m. until midnight, when I didn't have to stomp out fires and do the diplomatic dancing that fills up the days of most foreign agents. So finally, alone in the office, I would submerse myself in the line sheets, looking for that one piece of intel, that one critical clue in a torrent of misspelled and often barely literate Spanish text. My retinas burned as I sank deeper into Chapo's world.

EVERY DAY AT AROUND 11 A.M., Brady and his team in El Paso would see the key lieutenants, the offices, and the Second-Tier phones coming to life. This was the same modus operandi I'd seen among the traffickers I targeted in the United States. Diego and I would joke about "doper time"—drug dealers, whatever their level in the organization, are creatures of the night, waking up and conducting business whenever they're good and ready.

Brady and I were now witnessing firsthand the extent of Chapo's exploitation of new markets. Guzmán was eager to find refrigerated warehouses and place his operators in England, the Netherlands, the Philippines, and even Australia.

We knew, too, about Chapo's vast distribution network throughout the United States, but we were caught off guard by his deep infiltration of Canada. In terms of profit, Chapo was doing more cocaine business in Canada than in the United States. It was a straightforward price-point issue: retail cocaine on the streets of Los Angeles or Chicago sold for $25,000 per kilo, while in major Canadian cities it sold for upwards of $35,000 per kilo.

His key cartel lieutenants could exploit weaknesses in the Canadian system: the top-heavy structure of the Royal Canadian Mounted Police hampered law enforcement efforts for even the most routine drug arrest and prosecution.*

It was a perfect match for Chapo: hindered law enforcement and an insatiable Canadian appetite for high-grade coke. Over the years, the Sinaloa Cartel had built a formidable distribution structure, smuggling loads of cocaine across the Arizona border and hauling them to stash pads and warehouses in Tucson or Phoenix, before they were driven by car to the Washington border, where the loads would be thrown into private helicopters. The birds would

---

* Unlike the United States, whose federal law enforcement system comprises many specialized agencies—DEA, HSI, ATF, and FBI—Canada has only the RCMP, also known as Mounties.

jump the border and drop the coke out among the tall lodgepole pines of British Columbia.

Chapo's men had connections with sophisticated Iranian organized-crime gangs in Canada who were facilitating plane purchases, attempting to smuggle ton-quantity loads using GPS-guided parachutes, while sending boxes of PGP-encrypted smartphones south to Mexico at Chapo's request. A network of outlaw bikers—primarily Hells Angels—were also moving his cocaine overland and selling it to retail dealers throughout the country.

But Canada wasn't always smooth sailing for Chapo. At some point he'd entrusted a twenty-two-year-old from Culiacán who spoke decent English—Jesus Herrera Esperanza, a.k.a. "Hondo"—and sent him to Vancouver to run his drug distribution and money collection throughout Canada. Hondo's front—and it was a sweet life for a young Sinaloan—was to enroll in a business program at Columbia College, in downtown Vancouver, near his luxury thirtieth-floor condo loft. Hondo only attended a few classes, instead spending most of his time hanging out at clubs or taking girls sailing on the British Columbia coast.

But Hondo was sloppy and openly flaunted his connection to Guzmán. Brady and I hacked into Hondo's Facebook account one night and saw a status update reading:

Puro #701!

"What the hell is this kid posting?" Brady asked.

"Pure seven-oh-one?" Then suddenly it made sense. "It's not code, dude—it's *Forbes*." I laughed. "That's Chapo's *Forbes* number." Guzmán had recently been ranked by the magazine as the 701st-richest man on the planet.

Hondo was clearly a weak link among all the operators. He was so amped up about living the narco-junior life that he disregarded his daily functions for Chapo's DTO. At one point, millions of dollars were sitting uncollected in Vancouver, Calgary, Winnipeg, Toronto, and Montreal—all from the sale of Chapo's cocaine and heroin.

Finally, in frustration, Chapo—mirrored through Office-5—gave Hondo a direct order:

"I want a report every night at seven sharp. How much you've sold and how much money you're sitting on. Break it down by city."

When Hondo got around to sending in the numbers, we would read the nightly reports. Vancouver: $560,000 and 95 kilos of coke. Winnipeg: $275,000 and 48 kilos. Toronto: $2 million and 150 kilos . . .

I ALSO BEGAN to see how stuck in the details Guzmán could be.

In July 2013, a jerry-rigged go-fast panga, equipped with four Yamaha 350-horsepower outboard motors and 130 plastic containers full of fuel, had set sail from Ecuador with two young Mexicans at the helm. They had stashed their cargo in a fishing net: heavy garbage bags carrying 622 kilograms of cocaine. The men left the coast of Ecuador, charting a zigzag course, avoiding fishing vessels and coast guard patrols, sleeping in the open air, and eating only canned scallops and saltine crackers for a week at sea as they throttled northbound toward Mazatlán, in Sinaloa, Mexico.

They never made it. Tipped off that a Mexican navy vessel was headed out to intercept them, the two young men decided to ditch the load. A similar thing had happened to some of Chapo's other smugglers several months earlier; they, too, had nearly been intercepted, and had dumped their kilos of coke into the ocean, then had lit the remaining canisters of gas, turning their go-fast boat into a fireball as they jumped into the Pacific and nearly drowned. This time, the tip-off came early enough that the men had heaved the fishnet of waterproofed cocaine bricks overboard, attaching an orange buoy to it so that it could be spotted by air and retrieved.

Chapo was livid: losing one load was bad; losing two was unacceptable. His Mazatlán-based maritime smuggling lieutenant, Turbo, sent boat after boat out to the area of the drop, sixty miles off the coast, in a desperate attempt to find the lost load.

But we could tell Chapo was nearing the boiling point when he sent his best pilot, Araña, out in an old rattletrap Cessna to look for that bobbing buoy, making several flights a day, circling above the Pacific.

"That shit is probably in *China* by now," Araña complained to another pilot. "I can't take another day flying over the ocean. I'm fucking scared. The boss can ask me to do anything and I'll do it—but not this. I'm *not* flying out there again."

Brady and I couldn't believe how much manpower Chapo was putting into trying to retrieve a 622-kilogram load. It made no rational sense for the world's wealthiest drug kingpin to search so hard for 622 keys.

I HAD BEGUN to discern a psychological pattern in my prey: Chapo was fixated on the minutiae, like the price of jet fuel or the precise number of pesos being paid to his people. And he was cheap. For example, Guzmán would authorize monthly payments of only

2,000 Mexican pesos—roughly $165—to military lookouts scattered along the Colombia-Ecuador border. Why would he nickel-and-dime such key cogs in his machine of institutional bribery?

Chapo Guzmán was apparently the CEO of a sprawling, multi-billion-dollar drug-trafficking organization, but he also spent hours each day acting as a personnel complaint department. Brady and I couldn't help but laugh some days when we'd read the exchanges from Chapo's lieutenants bitching about being unappreciated—or, worse, not receiving their monthly payments on time.

I COULD ALSO GET stuck in my own obsessions, to the point of being oblivious to what was going on around me in the office.

One morning, I was so consumed with the line sheets that words began blurring together on the screen in front of me—wiggling back and forth. Was I having a panic attack? I looked over at the coatrack and realized that one of the hangers was rocking hard from side to side.

Terremoto.

Mexico City had frequent small tremors, but this was the first substantial earthquake I'd felt. After the massive earthquake in 1985—known to have killed more than ten thousand people—many new buildings were built to roll with the earth. The US embassy was built out of marble and placed on anti-seismic rollers for this purpose.

I DROVE HOME THAT EVENING thinking about a brand-new nickname I'd read about in Chapo's world. "Naris"—"the Nose"—was a courier who was constantly being sent by Second-Tier (once again through Office-6) to change cars frequently, pick people up, and deliver them to specific locations. Could he possibly be Chapo's personal gofer?

Narrowing in on Naris's location was now my new priority.

I parked my Tahoe on the street a block over from my apartment—something I did every so often to switch up my routine—and, walking home, I realized it was Día de los Muertos—the Day of the Dead, when Mexicans celebrate the deceased by dressing up in elaborate costumes and painting their faces as fanciful skulls with flowers and bright colors. The streets were full of people parading to celebrate in cemeteries, and my wife was throwing a party at home for all the women of the neighborhood, decorating sugar skulls she had made from scratch. She had quickly made friends with the extensive embassy and expat network; there were playdates with the kids and parties on all the major Mexican holidays like this one.

I smiled at the sight of my wife, taking full advantage of our time in-country, as I walked to the back bedroom to take off my suit and tie.

Unfortunately, there was no time for me to join in the festivities.

I sat down on the corner of the bed and quickly began reading a string of messages from Brady.

"How many times has Chapo been married?"

"No idea," I answered. "He's married at least four or five women that I know of. But no one knows for sure. He never divorces; he just marries again. Not to mention the women he doesn't marry. The guy is obsessed with women."

"I can see that," Brady wrote back. "Check this out. It just came over Second-Tier. Someone sent a photo spread of young girls in lingerie. Looks like they're coming from a madam they call Lizzy.

"He's sent a menu and gets to pick which one he wants for the night," Brady wrote. "What a fuckin' lowlife."

"Degenerate," I replied. "Sickening . . ."

Second-Tier then ordered Naris to the "Galerías" to pick up Lizzy's girl after Chapo had made his selection from the photo array.

I later determined that "Galerías" was code for the Centro Comercial Plaza Galerías San Miguel, a mall in the heart of Culiacán to which Chapo would send his visitors to rendezvous with Naris or other couriers to be brought to his secret hideouts.

NOT ONLY DID CHAPO have a fixation with underage virgins, but he had also become obsessed with the popular Mexican actress Kate del Castillo after becoming infatuated with the hit telenovela *La Reina del Sur*, on which she played a Sinaloa-born cartel boss running her empire from Spain. I had read in one line sheet that Chapo had instructed Lic Oro to get Kate's personal PIN so they could contact each other.

"This guy's got no other motive in life besides moving dope and banging as many women as he can," I wrote to Brady. "None. He's either obsessing over the day-to-day of the DTO or he's getting laid."

Sex was the only break in Chapo's workaholic drug-trafficking routine. He maintained a revolving door of women; in between, he'd invite his wife over to share the same bed; the sex was almost constant.

My street-cop instinct kicked in: the stronger the obsession, the more likely it could result in an exploitable weakness, a possible

Achilles' heel. I had even heard from a confidential source that Chapo and Mayo often joked that women would be their ultimate demise.

THE MORNING AFTER the Día de los Muertos party, I walked out to my armored Tahoe to find that the spare tire had been stolen. There was a small hole in my windshield—a circular shatter mark, spider-webbing the bulletproof glass near the driver's side. It looked like a close-range shot from a pistol.

I stepped back slowly, away from the Tahoe, my eyes darting from one car to the next, looking for anyone who might be surveil-ling me.

My eyes landed on someone sitting inside a black Lincoln Nav-igator across the street. Images of the scar-faced man I'd seen dur-ing the money drop flashed through my mind.

Could it be the same guy?

I wasn't about to stick around and find out. I quickly jumped into the Chevy, where at least I was protected by the armor. I called my wife and told her to stay inside for the day as I took off slowly, waiting for the Navigator to follow.

I hit the gas, took a sharp right, then another right, and quickly lost sight of the Lincoln in my rearview mirror.

AS I WALKED PAST the Kiki Room at the embassy, my phone rang.

"Hey, Tocallo has just asked Inge if he can have a guy killed in prison," Brady said. "He knows the guy's exact location—the cell he's housed in and everything."

We knew that Inge was short for *ingeniero*—"engineer"—yet

another nickname the DTO lieutenants and workers called Chapo over the BlackBerry messages.

"What's Chapo saying?

"It's weird," Brady said. "He's telling Tocallo to gather more information. He wants to know more."

That actually sounded liked Chapo. Despite his media-inflated reputation as a homicidal drug lord, I knew by now that Chapo seemed to be very deliberative, even circumspect, when authorizing the use of violence. In Sinaloa, most traffickers didn't think twice about killing someone, especially in the mountainous terrain of the Sierra Madre, where Chapo was raised—blood feuds and shooting wars were a simple fact of life there.

But Chapo must have become wiser over the years. Many times, when his lieutenants would report a serious problem—a killing offense—Chapo would conduct his own version of a police investigation, asking a series of questions to obtain more facts.

My mind flashed back to my days in the Phoenix Task Force when Diego and I would sit for hours with our assistant US attorney, drafting indictments, continually getting beaten over the head with questions, as if we were already under cross-examination on the witness stand:

"So *how* do you know this, Drew? Were *you* there? *Who* told you that?"

Chapo was exhaustive with his interrogations. He'd typically contemplate the best course of action for a day or two before making the calls to resolve the problem—even if the final outcome was a death sentence.

Brady and I confirmed this when we watched a video of Chapo taken several years ago, wearing his trademark plain dark baseball cap, casually walking back and forth underneath a palapa high in the Sierra Madre while an unidentified man sits on the ground with his hands tied to a post. Chapo's demeanor is calm and detached as he paces and interrogates the prisoner.

NOW THAT BRADY AND I were piecing together a good percentage of Chapo's life through the interception of Second-Tier, we needed to once again advance up the ladder of mirrors. Chapo's personal device couldn't be too far away at this point.

"Second-Tier is relaying everything up to a username labeled MD#8," Brady told me one day.

"Does MD#8 have a name?"

"Yes," Brady said. "Second-Tier's been calling him Condor."

I repeated the name over and over in my head, trying to remember whether I'd heard it before. Nothing registered. But unlike the usernames of Chapo's other mirror devices, "Condor" sounded like an actual person. Or at least like a narco nickname. Condors spend their time in mountainous country, soaring—was the name a clue that he was higher up in the cartel hierarchy? I couldn't waste my time speculating; I needed to know exactly where Condor was pinging.

I may have had Condor's PIN, but I still needed the corresponding Mexican telephone number to ping.

And for that I needed Don Dominguez. Don was a staff coordinator at DEA's Special Operations Division (SOD), in Chantilly, Virginia. The primary function of staff coordinators at SOD was to assist agents working high-profile cases in the field by coordinating deconfliction efforts, providing funding for wire intercepts, and acting as liaisons with the intelligence community.

Though equivalent in rank to my own group supervisor, at heart Dominguez was a street agent.

"He's not like the other desk jockeys in DC," I told Brady. "Don's one of us. He *gets* it. He believes we actually have a chance at capturing this fucker."

I sent Condor's PIN to Don so he could flip it. Don had access to a small team of techies at DEA, each of whom had built excellent relationships with the largest telecommunication service providers—even Canadian ones, like BlackBerry.

A standard request to a comms company could take almost three weeks to yield results, and by that time Condor—and all the other users—would be on to a brand-new BlackBerry and we would have to begin the process all over again. But once Brady and his team drafted an administrative subpoena to BlackBerry requesting Condor's corresponding telephone number, I was confident that Don would work around the clock to get the subscriber results back quickly.

SURE ENOUGH, IN LESS than twenty-four hours, Don Dominguez delivered.

"Just got Condor's number back from Don," I told Brady, anxious to hit the ping button on my laptop.

"Where's it at?" asked Brady.

Within minutes my eyes lit up as I received the results and sent the coordinates back to Brady:

24.776,-107.415

"It's hitting in Colonia Libertad."

"Colonia Libertad?"

"Yes," I said. "Looks like a small, run-down neighborhood on Culiacán's southwest side."

Now we had a BlackBerry in the heart of Sinaloa's capital. The net was narrowing: it was the first ping we'd ever had outside of Durango.

"CONDOR" WAS IN CULIACÁN. An average-size city of 675,000 nestled in the center of Sinaloa, just west of the Sierra Madre, Culiacán is the birthplace of all Mexican drug trafficking and had displaced Medellín, Colombia, as the world's narco capital. From the days of Miguel Ángel Félix Gallardo through to Chapo's current reign, all the top cartel leaders had come from the city or the small towns nearby.

Culiacán was also famous for its Jardines del Humaya cemetery—the "drug lord's burial ground"—with its $600,000 air-conditioned mausoleums, including a lavish marble one for Chapo's murdered son, Edgar, and a large shrine to Jesús Malverde, the mustached patron saint of drug trafficking. Legend had it that Malverde was a bandit from the hills of Sinaloa who stole from the rich and gave to the poor until his death by hanging in 1909.

I REMEMBERED DIEGO telling me about visiting Jardines del Humaya when he was once on vacation in Culiacán. Diego said he was astonished by how much money traffickers had poured into the shrine to keep it thriving. Now Culiacán was known as a city of outlaws, and off-limits to authorities from outside of Sinaloa, which was a problem, because most of the local cops and military had been corrupted by Chapo's organization.

In fact, no outside law enforcement or military personnel had ever dared to enter Culiacán to conduct an operation, for fear of immediate retaliation.

Still, as distant and untouchable as Culiacán seemed to me, this was our first indication that Chapo could be in Mexico's narco capital.

I QUICKLY GRABBED the next flight from DF to El Paso and met with Brady, Joe, and Neil Miller, the other member of Brady's core team at HSI.

"Neil's our bulldozer," Brady said, laughing. "He doesn't think twice about pissing someone off, as long as the job gets done. Welcome to his domain."

Brady shoved open a door to reveal their newly created war room, a converted conference area discreetly tucked away from everyone else at the HSI El Paso Field Office. They'd recently taken over the room and filled it with more than a dozen computers and at least that many translators, to run all the wires on the office devices, on Chapo's key lieutenants, and now on Second-Tier.

But despite the resources, Brady was still on edge. "How do we know that there aren't a hundred more layers to the pyramid-like Second-Tier and the offices? I think we're fucked. The mirrors could go on forever."

I paused for a second before giving my partner the breakthrough.

"No, there aren't *hundreds*," I said. "I've been analyzing Condor's tolls, looking at his most frequent contact. I just found it: two. There's only *two* layers."

Condor wasn't in contact with any other PINs—just Second-Tier.

"It stops right there," I said. "Condor isn't forwarding any messages. He's the end of the line."

Brady couldn't believe it.

"Condor is either fat-fingering thousands of messages a day into a new BlackBerry and forwarding them on—a nearly inconceivable job—or he is standing in the same room as Chapo, receiving personally dictated orders directly from the boss," I said.

Brady rushed into the wire room and returned several minutes later with Neil.

"We've *got* him, brother," said Brady.

"What do you mean?"

"I'm looking at it right here."

Brady showed me a line sheet that had come in that morning from Second-Tier to Condor, asking if *el generente*—the manager—was awake yet.

Condor clearly was in the same house—or even the same room—as the boss.

Condor had replied:

"No, he's still sleeping . . ."

ABRA LA PUERTA

TOP-TIER HAD BECOME MY LIFE.

Pinging that BlackBerry device, the one nearest to Chapo, was all-consuming. As long as I could ping Top-Tier—from six in the morning often until after midnight—nothing else really mattered. Even when lying in bed with my wife in La Condesa, my mind was never far from hunting that Top-Tier.

By now I knew *how* Chapo ran the day-to-day of his multi-million-dollar drug empire; all I needed was the boss's location. This wasn't as simple as it sounded, given Chapo's penchant for bouncing constantly, moving from safe house to safe house, from countryside to city, sometimes on an hourly basis. With every ping, I meticulously labeled the spot with a yellow thumbtack on my Google Map, marking the coordinates along with the date and time indicating where Condor's device had pinged in Culiacán.

Top-Tier.

If Condor was standing with the man, every new ping helped me begin to establish Chapo's pattern of life.*

---

* "Pattern of life" is the investigative term of art for a target's location history up to the present moment.

Brady, Neil, and Joe were working around the clock now, too, intercepting as many mirror devices as they could identify— Offices 1 through 10, and Second-Tier—as well as a new critical mirror who went by the username "Usacell." We quickly determined that Usacell—similar to the name of another major Mexican telecom service provider Iusacell—was a duplicate: another Second-Tier device run by the user Telcel, in Durango.

"It's pretty obvious it's the same guy," Brady said. "He's just labeled each of his two BlackBerrys with the corresponding service provider to tell them apart."

The Usacell device may have been another mirror, but it exposed still more important messages that Chapo thought were hidden. If the office devices were sending two hundred messages a day to Telcel at Second-Tier, they were sending an equal amount to Usacell. Brady and I estimated that we were intercepting close to seventy-five percent of all the DTO communications coming to and from the boss.

The window into Chapo's world was now becoming brighter.

"For now, we should sit at Second-Tier," Brady said.

At the Second-Tier level, we could intercept every order coming down from Chapo and every communication coming up from the office devices.

"Yeah, that's definitely the honey hole," I said.

If Condor and the offices dropped their BlackBerrys, Brady and I could identify their new PINs easily, so long as we were still intercepting the two Second-Tier devices, Telcel and Usacell.

Office-4 was now starting to produce valuable intel, too, but I noticed something different about this mirror: not only did Office-4 appear to be sending messages up the chain to Chapo through Second-Tier, but it was also responsible for relaying command-and-control messages—mostly related to Chapo's Canada operations—to another top player who went by the username "Panchito."

"Did you see the deconfliction hit on Panchito?" I asked Brady. "It's hitting all over FBI New York."

"Yeah," Brady said. "I saw it."

"Our Panchito has got to be Alex Cifuentes," I said.

The FBI New York office claimed to still have an interest in Chapo after it began targeting him through longtime Colombian drug lord Hildebrando Alexánder Cifuentes Villa, who'd moved to Sinaloa around 2008—acting as human collateral for all of Chapo's cocaine shipments generated by the Cifuentes-Villa family in Medellín.

After the failure of the Cabo op, the FBI's fresh intelligence slowly dried up. Alex—as everyone called Cifuentes—was one of Chapo's right-hand men.

In fact, months prior to my Mexico City coordination meeting, while I was in New York, I'd sat down with the FBI and told them about the great working relationship I was building with HSI and Brady's team.

"We're moving quickly," I said. "This train isn't stopping. If you guys want to get on board and share your intel, now's the time."

This wasn't my first attempt to coordinate a joint investigation with the FBI. I'd found their special agents to be polite and professional, but I also knew they were highly resistant to sharing. It was typical of the FBI to hold their cards close to their chest: that's how they were trained at Quantico. The FBI believed they were the world's premier law enforcement agency, but when it came to working a drug investigation—especially when faced with the complex structure of the Mexican cartels—their expertise couldn't match that of the DEA.

As much as I tried to get everyone to cooperate, I knew it was going to be difficult.

The FBI's file was composed mostly of historical intelligence on Cifuentes, who was now wanted by DEA and FBI after being federally indicted on multiple drug-trafficking conspiracy charges. But instead of sharing with the DEA, the FBI began giving their intelligence to the CIA, in hopes they could produce something that would give them the upper hand.

I knew that whenever intelligence was passed to CIA by a federal law enforcement agency, the source would instantly lose control of how that intelligence was classified, disseminated, and used. This was well known by the agents who worked in the embassy, and it was precisely why Brady and I had decided that the CIA had no place in our investigation.

Almost every piece of intelligence we gained on Chapo was derived judicially from court-authorized wire intercepts, so that the evidence collected could be used to charge Chapo and others in his DTO in a US federal court. It was exactly how DEA disrupted and ultimately dismantled DTOs. The CIA, on the other hand, dealt extensively with classified and top-secret material that was difficult—if not impossible—to present in court.

I didn't need the CIA, but I also knew that they were anxious to get involved now that Brady and I were gaining momentum toward Chapo's exact location.

"The Feebies and the spooks want to call a meeting," I told Brady.

"Where?"

"Langley."

"Fuck that," Brady said. "We don't need them."

"We need to at least be on the same page when it comes to Cifuentes. We need to send someone if you or I don't go. I'm going to talk to Don."

Don Dominguez had been following these developments from Virginia and agreed to attend the meeting at CIA headquarters on our behalf. The result of the meeting was an agreement among all agencies to arrest Cifuentes and remove him from Chapo's DTO, but *only* at the right time. It was crucial that the efforts be coordinated among all agencies. I confirmed with the FBI that Panchito's PIN was in fact Alex Cifuentes and shared several of the ping coordinates I had obtained from the Cifuentes BlackBerry, hitting as it did in a rural area just southwest of Culiacán.

IN LATE NOVEMBER 2013, I received an urgent text from Brady in El Paso.

"This just in," Brady wrote, quoting the line sheets after the Spanish translation. It was Second-Tier transmitting to all the office devices:

"Panchito was caught in a battle with soldiers and Picudo went to rescue him. Turn off your phones because they will get your PIN."

I called Brady immediately.

"Goddammit—the Feebs fucked us!" he shouted.

"Hold on," I said. "Let me look into it and get the facts."

I reached out to DEA Mazatlán, who in turn contacted their local military contacts to see if they'd heard about a recent arrest just outside Culiacán.

Initially, the Mexicans didn't even know *who* they'd arrested. SEDENA* had locked up some middle-aged guy at a small ranch, but they didn't think he was Colombian, and his name wasn't Cifuentes.

"They're saying they've got a guy called Enrique García Rodríguez," I told Brady. "They're getting me a photo of him right now, along with the passport."

Brady stayed on the line while I waited for Mazatlán to shoot me the email.

When the photo arrived, it showed a man in his mid-forties, with a receding hairline, salt-and-pepper beard, and light complexion.

"It's Cifuentes, man," I said. "It's a fake name on this Mexican passport. Panchito is done."

"Fuck them!" Brady was livid.

He knew it was just a matter of time before everything we'd built in the war room in El Paso came crashing down.

Sure enough, within minutes, Chapo's offices were already talking about dropping their BlackBerrys; Second-Tier wouldn't be far behind.

And then Top-Tier: Condor.

Brady and I would soon be standing, once again, in the dark.

"I just confirmed the photo with FBI," I said. "They're claiming they had nothing to do with this."

"Bullshit," said Brady.

"I don't know for sure. But I can promise you—the CIA are the ones who gave the information to SEDENA," I said. "Guaranteed."

After I got off the phone with Brady, I reached out to the CIA's

---

* The Mexican army—short for Secretaría de la Defensa Nacional.

counter-narcotics group in Mexico City about the Cifuentes arrest. At first they denied any knowledge, but a few days later, a CIA manager told me the truth: all the rural ping locations I had shared with the FBI had been passed on to SEDENA by the spooks. (The CIA claimed to have told SEDENA only that there was a "subject of interest" at that location.) The CIA then turned around and washed their hands of the whole ordeal after passing the lead— there was no oversight of the operation, and no close coordination with their Mexican counterparts. In fact, the CIA didn't even know if SEDENA had captured the right man, otherwise they would have claimed credit immediately. Once I confirmed that it was Alex Cifuentes, though, the CIA were happy to step up and take credit.

I was sickened by the CIA, but I also knew this was simply the way they operated: collecting intelligence from their own government and then haphazardly sharing it with their Mexican counterparts.

"That's the fuckin' spooks, man," I told Brady. "They tell the Mexicans—then just stand back and watch the shit show. The CIA doesn't give a fuck about dismantling international DTOs. It's just another stat to them. If they can stat the pass-on of key intelligence or an arrest, then they can justify their existence."

It was a classic example of the breakdown in communication, if not outright antagonism, between the US intelligence community and federal law enforcement. I knew the ropes by now: most CIA activity in Mexico was never coordinated or deconflicted with the DEA. It was true of their approach in Mexico, but it was also how the CIA operated around the rest of the world. They often caused major disruptions to highly sensitive, judicially authorized investigations like ours.

Whatever the case, Alex Cifuentes had been prematurely arrested by SEDENA, and we were left to pick up the pieces. Virtually every BlackBerry we'd been intercepting dropped the day after the Cifuentes arrest. I was pissed, but I held back my emotions when speaking with Brady—no sense adding more fuel to the fire. It didn't stop him, however.

"I'm fuckin' done with the FBI," Brady said. "We're not sharing a single piece of intel with them ever again."

"I understand," I said. "But we need to keep an even keel here. The thing is—I don't know what else CIA may have from the FBI that could torpedo this thing even further. We've got to keep them close."

"All right, dude," Brady said. "You keep playing Switzerland— it's what you do best. Listen, if I was there in Mexico right now, I'd be choking some motherfuckers out."

"ARGO," I SAID. "You see that movie? Ben Affleck?"

"About the Hollywood sting with the Iranians?"

"Uh-huh."

"Sure. 'Argo fuck yourself.'"

I laughed. "Think we could pull it off?"

"With Chapo? I don't know . . ."

For years, Alex Cifuentes had been searching for producers, screenwriters, and authors—all at the specific request of Chapo.

Strange as it sounded—given the precautions he took to guard his location and the secrecy of his communications—Chapo had become fixated on telling his rags-to-riches life story. He was desperate to see his rise from that impoverished little kid selling oranges in La Tuna to the world's wealthiest drug lord on the big screen. In the line sheets, we would sometimes read about how Chapo was batting around the idea of a film, telenovela, or book. Chapo would entertain just about *anyone* who was interested in hearing his story.

Accordingly, Alex Cifuentes would get recommendations through his contacts for various filmmakers and writers and then vet them for Chapo. If they passed muster, Cifuentes would schedule a face-to-face meeting with Guzmán at a secure location, somewhere in Culiacán or at a ranch in the mountains.

Brady and I had learned of at least one aspiring filmmaker—we knew him only as Carlino—who'd flown in from Cabo. Carlino had Hollywood connections and claimed to have worked with the producers of the hit Fox TV show *Cops*.

Chapo knew the show *Cops* and was very interested in following up.

"We need to reach out to them," said Brady.

"I bet they'd be on board," I said. "I've already spoken to my guys in Los Angeles. They have a few DEA agents with connections to producers. They'd be willing to work with us."

"Set up our own version of *Argo*?"

"Exactly."

"How would we play it?" Brady asked.

"You could be UC as the director," I said. "Just get one cameraman working with us as an undercover. You'd be perfect, dude. Keep your usual poker face. Never smile. Throw on some horn-rim glasses and just grumble and curse at everyone."

"I could handle that," Brady said.

"Let's get up to LA to see what our options are," I said. "With Cifuentes locked up, finding the right undercover agent to pose as a producer or screenwriter may be our only way to get this *cabrón* out of Culiacán."

"We could call the film *Saludos a Generente*!" Brady said.

"Hell, no. I've seen that line one too many times already."

I was staring at a map of Mexico, looking at all the possible coastal resort towns for a meet.

"The beauty of it is," I said, "Chapo wouldn't even have to leave the country. He'd go back to Cabo if it was to start rolling footage for a movie based on his life. Vallarta? Even Cancún would probably work. Anywhere on the Mayan Riviera. He has a history of traveling over there already. He'd feel safe."

NOT ONLY WAS CHAPO conducting meetings with producers and writers, but I'd learned about a thumb drive on which he had the first half of a movie script about his rise to power. He'd let his wife Griselda López Pérez and daughter Grisel Guzmán López review it, only to have them complain that the screenplay didn't mention them enough.

Griselda, Chapo's second wife, was accorded a special level of respect—even deference—by the drug lord. Brady and I had intercepted Griselda's complaints: she often demanded more money for her kids, and Chapo would comply, handing over $10,000 every few weeks.

Despite being exes, they clearly were still close. Guzmán and Griselda had three surviving children—Joaquin, Grisel, and Ovidio—and they were some of Chapo's favorites.

For months, Brady and I had been intercepting Joaquin and Ovidio. The brothers were going by the code names "Güero" and "Ratón."

Being light-skinned, Joaquin's use of the name Güero was an obvious choice.

I remembered the first time Diego had taught me the word *güero*, years back when we were listening to narcocorridos during our Phoenix Task Force days. "You come down to DF with me," Diego had said. "Pick up some slang, eat the street food—they won't take you for a gringo, dude. Everyone'll call you *güero* . . ."

"And why Ratón?" Brady asked.

I had studied the one photo I had of Ovidio. "He looks like a mouse," I said, which cracked us both up. The kid did have big black eyes and protruding ears . . .

Güero and Ratón were speaking constantly with their father through the mirror of Office-1, just like their half brother Iván.

The two sets of brothers operated in pairs, but Güero and Ratón seemed to be more heavily involved in Chapo's day-to-day business than Iván and Alfredo. According to the messages we were intercepting, all four sons were key players in Chapo's drug dynasty, though, and closer to him than anyone else in his organization.

These guys weren't wannabes: the hard-core narco life was in their blood—they had followed in their father's footsteps from an early age. Just from watching their day-to-day communications, I knew the four boys meant everything to Chapo.

A FEW WEEKS AFTER the arrest of Alex Cifuentes, Brady, Joe, and Neil, with the help of their lead prosecutor, Camila Defusio, had finally righted the ship in El Paso; they had cranked out the affidavits and were back live again, sending fresh intercepts down to me by the hundreds.

HSI was moving at lightning speed—all due to the management in El Paso providing full support for Brady and his team.

Brady told me that this was the biggest drug investigation HSI had ever been involved in. I knew that Brady's bosses were fully invested in their success and had been greasing all the logistics behind the scenes for months. I had never seen anything like it. Brady's brass kept us moving forward without the slightest bureaucratic interruption. It was impressive.

Brady and his team had climbed back up and were now intercepting a handful of office phones and a new Second-Tier device. But these grunt workers in Durango—however valuable the intercepts—didn't get us any closer to Chapo himself. Only pinging Top-Tier could do that.

I HAD BEEN WAITING to find the new Top-Tier device, and fortunately it didn't take us long. This time the username was MI-26, and in the line sheets everyone was calling him "Chaneke."

"Who's this now?" Brady asked. "What happened to Condor?"

"I'm not sure, man."

"Who the hell is Chaneke?"

"For now," I said, "let's assume he's our new Condor."

I pinged the phone on my laptop. "Perfect. Chaneke's device is hitting in that same neighborhood. Right in Colonia Libertad. That's where I had Condor."

I quickly Googled the name Chaneke. Like so many of the words in the line sheets, it turned out to be a phonetic misspelling. *Chaneques* were, in fact, among the hundreds of gods and spirits sacred to the ancient Aztecs. Legendary creatures in Mexican folklore, they are the "little people who steal your soul." The images of *chaneques* that I found—pre-Columbian sculptures and drawings—resembled tiny trolls with oversize eyes. By Aztec tradition, the *chaneques* were guardians of the forest, attacking intruders, frightening them so that their souls would abandon their bodies.

This Top-Tier Chaneke was also a kind of guardian: Chapo's direct intermediary. In the intercepts, the workers kept referring to Chaneke as "Secre."

"It's got to mean he's Chapo's secretary," I said. "And these usernames MD-8 and now MI-26—I think they're makes of helicopters."

"Maybe Chaneke's a helicopter pilot?" Brady said.

The reason, for now at least, remained a mystery, but one thing was clear: Condor might run a tight ship, but Chaneke was intensely disliked by the workers in Chapo's organization.

"Everyone *hates* Chaneke," Brady said, "Second-Tier and the offices keep bitching about him. Sounds like he's stiffing them. They're constantly asking him when El Señor is going to pay them. One office worker has been complaining that he needs the money to buy groceries for his kid. And Second-Tier told him, 'Don't worry, Condor will take care of you when he returns.'"

"I see it now. Two *secretarios*—Condor and Chaneke. Same job, they're just taking shifts," I said.

We determined that each secretary was working fifteen to thirty days straight—no downtime at all. They probably ate whatever the boss ate, slept whenever he slept, and typed out every order and whim Chapo needed sent via that Top-Tier BlackBerry.

"Talk about a 'pattern of life'—can you imagine Condor's?" I asked Brady. "Guy's got no personal time at all. Twenty-four seven he's Chapo's slave."

Brady let out a short laugh. "And the poor bastard gets to sleep in the room next to Chapo—listening to him bang his whores all night long."

We had figured out that Chapo, still spooked by the near capture in Cabo, no longer used a phone. He was strictly dictating orders now—his two assigned secretaries would relay all of his communications so that he didn't even have to touch the Black-Berrys.

BRADY AND I HAD our own reasons for disliking Chaneke. Whenever he took over from Condor, shift-change protocol was followed with almost military precision, meaning that Chaneke, the Second-Tier, and office devices in Durango would all drop their phones almost immediately, either tossing them in the trash or giving them to a family member to use. Either way, it created an instant logistical nightmare for us. Whatever phones we were intercepting—usually hundreds—all went dead. In an instant, no more messages, and no more line sheets to decipher. Our keyhole into Chapo's secret inner world would vanish, just like the ghostly drug lord himself.

This emotional roller coaster of being up, listening, one day, then down the next was beginning to grind on everyone's nerves.

"I don't know how much more my team can take," Brady told me after Chaneke had once again burned all the phones instantly.

"We're right there," I said. "Hold tight, brother. This last ping was in the same area in Culiacán."

I WAS CERTAIN THAT zeroing in on either Condor's or Chaneke's device would lead us straight to Chapo, so for fourteen hours straight I kept pinging. By now I had a concentration of yellow pin markers on my Google Map—adding to weeks of pings already and forming a swirling bull's-eye in the heart of urban Culiacán. I was staring at a clear pattern of life. Although there was some slight variation in the pings, it appeared that Condor and Chaneke never moved from that Colonia Libertad section of Culiacán.

"Have you seen the overhead images on these?" I asked Brady.

"Looks like a goddamn shantytown down there," Brady said. "Dirt streets. Derelict cars. Tarps draped over back lots."

"Yeah, the shitty neighborhood doesn't make much sense. Why would Chapo be holed up there? We know his secretaries need to be face-to-face. Right there with him. Maybe he's laying low."

"Hold on—these are just coming in," Brady said. He had a batch of brand-new translated line sheets. "Chaneke instructed Naris to go pick up Turbo at the Walmart 68. Naris is driving a black Jetta."

"Turbo's in some serous shit," I said.

"Those six hundred kilos are long gone, and Turbo's personally responsible," Brady said.

"Yeah, could get ugly. Turbo will be lucky if he makes it out alive."

Several hours prior, Chaneke—passing orders down through Second-Tier, then to Office-5, and finally out to Chapo's Mazatlán-based maritime coordinator—instructed Turbo to head to "El 19." His *tío* ("uncle," yet another coded nickname for Chapo) was ready to see him. Turbo was then told to go to the "Walmart 68 on Obregon," where he'd wait to get picked up by a black Jetta.

Brady and I had established that El 19 was clearly the DTO's code for Culiacán.

"And I know that Walmart," I said. "It's just to the east of all my Top-Tier pings, off Avenida Álvaro Obregón. It's close—the perfect pickup spot." I watched the movements of Naris as I continued to ping him.

"You dialed in?" Brady asked.

"Yep, he's there at the Walmart. Now he's headed further west."

I kept tracking Naris with regular pings, getting closer to the elusive boss.

Abra la puerta. Abra la puerta!

Naris was sending messages out to Office-6, but there was a glitch in the system. His requests to "open the door" weren't getting delivered to Second-Tier, which meant they weren't getting relayed to Chaneke.

"He's stuck there at some front gate," Brady said. "Frustrated as hell."

It was a bizarre turn of events. I knew Naris was usually highly cautious about all communications—he'd almost always power down his BlackBerry before getting close to Chapo's location. But now Naris was stuck outside the house, and with no one opening the gate for him, he'd decided to risk it, turning his BlackBerry back on to push his messages out quickly.

Carnal, abra la puerta!

"That's it!" I said.

Naris was now pinging directly on top of Chaneke.

"Yeah, Naris is at the gate," I said. "We've got him, Brady. Not sure which house. But Chapo is definitely right there on that fucking block."

DUCK DYNASTY

FOR MONTHS, BRADY HAD been asking me who we could trust among the Mexican counterparts to launch a Chapo capture mission.

"The Federal Police?"

"No—out of the question."

"Any units in SEDENA?"

"Not a chance."

But now, with our near certainty that Chapo was living within that one-block radius on my map, the answer became clear.

"There's only one real option," I said.

Only one institution within the Mexican government had a reputation for being incorruptible: the Secretaría de Marina–Armada de México (SEMAR).

"The marines?" Brady said.

"Yeah," I said. "Among the counterparts, SEMAR's all we've got . . ."

"You trust them?"

"Can't say I trust them," I said. "I can't say I trust *anyone* here. I've never worked with them, but I know they're fast and lean and always ready for a fight."

I'd been studying SEMAR's track record during their work

with other DEA agents at the embassy—they'd helped decimate the Gulf Cartel and then the Zetas, on the east coast of Mexico.

"Sounds promising," Brady said.

"There's a special brigade here in Mexico City," I said. "From what I hear they're the least corrupt of them all."

Since I began working with Brady and his team, I had shared *nothing* with any Mexican counterparts.

First, the Mexicans didn't yet know that US federal law enforcement was able to intercept BlackBerry PIN messages between two traffickers in Mexico. Second, there was no way I would release any intelligence prematurely without having Chapo's location pinpointed and ironclad.

"It's still too early to approach them," I said to Brady. "And I'm being told by the bosses here that SEMAR won't even consider going into Culiacán. Far too dangerous."

MEANWHILE, A NEW NAME had appeared suddenly in the line sheets.

"Lic-F," Brady said. "Have you seen this guy? I keep going over his messages. He's obviously very close to Chapo—looks like he's helping coordinate coke loads in and out of Culiacán, and he's very tight with Picudo."

"Yeah," I said. "Seems to me he's Chapo's most trusted set of eyes and ears. He's cautious and smart. But I don't think he's a lawyer. Some of these line sheets make me suspect he's actually got a law enforcement background."

The thought of Lic-F took me back to that escape from Puente Grande prison—Chapo's corruption of the guards, and even the failures of the prison management. Lic-F? El Licenciado? Dámaso? The former police officer in the Sinaloa Attorney General's Office who'd become a close friend of Chapo's during his stint in Puente Grande?

"I think Lic-F's going to turn out to be Dámaso López Núñez," I said. "But I can't tell yet. The only thing I'm sure of is that this guy is slick. And he's got some serious hooks within the government."

"Look at this," Brady said. "He's giving Chapo the status on a tunnel."

I pulled up the line sheet from Lic-F to Top-Tier. Lic-F was providing Chapo with a precise description of a tunnel that had been under construction for more than a year. "It's going to measure approximately eleven hundred meters, and they have finished more than six hundred meters," I read in the translated message. And Lic-F said he'd need less than a "roll"—$10,000—to finish construction and keep paying the salary of the tunnel workers.

"Damn," I told Brady. "That tunnel is gonna be more than a quarter-mile long."

"They're digging into San Diego or Nogales—one of the two," Brady said.

Working in the HSI office in El Paso, Brady had become an expert on tunnels along the US–Mexico border.

It was Chapo who had pioneered the narco tunnel along his key smuggling corridors. The tunneling had started nearly a quarter-century earlier, in 1990, when the first cross-border one was found in Douglas, Arizona. The Douglas tunnel, estimated to have cost the traffickers $1.5 million, originated inside a house in the town of Agua Prieta, Sonora, and ended some three hundred feet away at a warehouse in Douglas. Used for smuggling loads of weed for the Sinaloa Cartel, the media had dubbed it the "James Bond Tunnel," because the only way to access the underground passage was to turn on an outdoor water spigot at the Agua Prieta house, which triggered a hydraulic system that raised a pool table in a game room, which in turn exposed a ladder down.

No one knew exactly how many tunnels Chapo had constructed in the ensuing years. Since that first Douglas discovery, US authorities had found more than 150, almost always with the same signature construction features: ventilation, lighting, sometimes rails, and often sophisticated hydraulics were included.

I HAD DISCOVERED a key player in the tunnel construction, too: "Kava."

"Kava may be an architect," I told Brady. "Possibly an engineer. He's constantly reporting to Chapo, giving him status updates on his workers and the various projects they having going on. One job is up in Tijuana—probably the tunnel that Lic-F is talking about."

"Could be," Brady said. "Everything I'm seeing that's tunnel-related I'm passing to my guys in San Diego and Nogales."

On October 31, 2013, the San Diego Tunnel Task Force, comprising DEA, HSI, and US Customs and Border Protection, discovered a major tunnel between a warehouse in Tijuana and another in San Diego. I followed the news coverage live on CNN in the embassy.

This "super tunnel," as authorities dubbed it, went down to a depth of thirty-five feet and zigzagged about a third of a mile until the exit in an industrial park west of the Otay Mesa port of entry. Muling drugs down there must have been claustrophobic work; the tunnel wasn't big enough for a man to stand up in—only four-feet tall and three feet wide—but it did have ventilation, lighting, hydraulic doors, and an electric rail system.

US Immigration and Customs Enforcement locked up three suspects and made a sizable seizure during the discovery of the super tunnel—more than eight tons of marijuana and 325 pounds of cocaine.

Brady and I suspected that the super tunnel was Chapo's handiwork—not because of chatter in the line sheets but because, after its discovery, the DTO's BlackBerrys immediately went completely silent about it.

"Amazing," Brady said. "Everyone's quiet. They're disciplined. They lose a tunnel of that size and no one's saying a goddamn word?"

"With as many tunneling projects as Kava's working on, Chapo's gotta be used to these things getting popped by now," I said. "Hell, they've probably got at least another five super tunnels in the works."

AS BRADY AND I were in the heat of strategizing our high-risk operation to capture Chapo, the Mexico City embassy was rocked by shocking news. In mid-December 2013, a joint capture-op involving units of the DEA and Mexican Federal Police in the beachside resort of Puerto Peñasco, just south of the Arizona border, turned wildly violent.

I had just finished breakfast and was fixing the Windsor knot in my tie when the phone rang. "Get in here *now*," my group supervisor said. "Marco and the guys are pinned down up in Sonora in a shoot-out. They're calling for help."

I quickly snatched my laptop bag and headed toward my G-ride. On most days, DEA Special Agent Marco Perez sat next to me, but on this particular morning, Perez, several other DEA agents, and the Mexican Federal Police were staging a covert operation in Puerto Peñasco to arrest Gonzalo Inzunza Inzunza, a.k.a. "Macho Prieto," a high-ranking leader in the Sinaloa Cartel. Macho Prieto ran his own drug-trafficking organization under the cartel's umbrella. A protégé of Ismael "Mayo" Zambada García, Macho Prieto was considered extremely violent.

I burst into the embassy hoping to hear Marco and his team were out of harm's way, and quickly got briefed on what had gone wrong. The Federal Police had approached the door of Macho's condo in the predawn darkness, and Macho's response had been instant. The cops began taking fire through the front door from Macho and his bodyguards, and within seconds there was a firefight under way in the middle of an upscale residential neighborhood filled with American tourists, just steps from the white sand. Macho's men—armed with AK-47s and automatic belt-fed machine guns—fired until they ran out of ammo.

Macho called in reinforcements, and gunmen raced in from other condos, firing at the cops from balconies and vehicles. Macho's "war wagon"—an armored white Ford F-150 with a 50-caliber machine gun mounted in the back bed underneath a camper shell—came screaming in through the front gates of the complex, smashing PF's barricade of cars, while snipers fired round after round into the front windshield, wounding the driver. The war wagon sped out of control, leaking gasoline and oil onto the pavement. Gunmen jumped from the front and back of the truck and ran to join the fight.

"They're in the back of the complex, hunkered down," my supervisor said, her ear to a phone receiver. "I can hear the gunshots over the phone in the background—it's nonstop." Perez and the other two DEA agents were around in back of the complex, taking cover behind a small concrete wall. The American agents were pinned down in the darkness and couldn't leave their positions because PF had two Black Hawk helicopters in the air, firing grenades at the bad guys, turning vehicles into fireballs.

Even local police cars began responding, but not to join the good guys—the local cops, all on Macho's payroll, were picking up the injured cartel gunmen and taking them away like a makeshift ambulance service. The PF team was small, outgunned, and now running the risk of being surrounded by Macho's men.

In the chaos, Macho's bodyguards were able to drag him out the back door and into a car to make an escape, but Macho was bleeding profusely. When PF finally made entry into the condo, they saw puddles of blood and scarlet handprints everywhere. Macho had sunken into a hot tub in an attempt to control his bleeding, turning the bubbling water thick and dark red. The smears continued on the floors and out the back door. Macho Prieto had escaped the gunfight but would soon die from his wounds.

Two Mexican Federal Police were also badly wounded; the DEA team raced them in a convoy across the US border to Tucson. They couldn't risk staying in Mexico a moment longer, for fear of being attacked—nowhere in Sonora was safe once word was out that PF had killed Macho.

The wounded PF members all recovered in an Arizona hospital; no DEA agents had been injured; Macho was dead—so the operation was judged a major success against the Sinaloa Cartel, but it was also one of deadliest international operations DEA had ever conducted.

CULIACÁN HAD ALWAYS weighed heavy on my mind. Even more so after the bloody operation to capture Macho Prieto.

Just as Macho maintained ironfisted control of his turf in Sonora, the city of Culiacán was in the grip of Chapo. Brady and I knew it would be nearly impossible to hit Chapo anywhere in Culiacán: we could end up in a firefight with an entire city. If Macho could summon that many fighters in the little resort town of

Puerto Peñasco, how many would Chapo have streaming to his aid if we were to hit him in Culiacán? It was precisely why Brady and I were working so hard to find a location outside the city where we could nab him quickly, quietly, and hopefully without a gun battle.

"How many kids does Chapo have?"

It was our fifth phone call of the day, and it wasn't even noon.

"All the women he's run through—no one really knows," I said. "Hundreds? You might be living next door to one." Brady let out a loud laugh.

"But really," I continued, "there's only the four key sons we need to pay attention to."

With Christmas season approaching, I saw that Ratón and Güero were taking regular trips out of Culiacán; Chapo would tell them where to meet—some place he called "Pichis."

"This new 'Pichis'—over and over. Just today again: 'Meet me at Pichis,'" Brady read from a freshly translated line sheet.

I had been dissecting those sheets as well.

"Yes, I've seen that. No idea what Pichis means. But he's building a pool there along with some palapas—Kava's been sending him regular updates on the construction."

Being hypervigilant for references to meetings among all the players was a top priority for me, especially if the get-togethers were planned outside the city. I started pinging Güero's and Ratón's phones simultaneously as they headed south from Culiacán; then more pinging, until I had six red pushpins tacked to my Google Map, tracing a crooked line down Sinaloa State Highway 5.

But then, nothing. After about fifty minutes, the sons' phones would stop pinging. Maybe they were so far in the sticks that they were out of reach of a cell tower? Or had they shut their phones off, or taken out the batteries, right before a meeting?

Chapo Guzmán after his first arrest in June 1993.

AP Photo/Damian Dovarganes

Boarding the DEA Learjet en route to Mexico City in June 2010, carrying the $1.2 million in cash inside FedEx boxes.

Diego sitting with Mercedes Chavez-Villalobos and her associates at La Rosita restaurant in Panama, June 2009. I took this photo surreptitiously during the undercover operation.

With the seized 2,513 kilograms of cocaine in Guayaquil, Ecuador, in November 2010.

Chapo in his signature black baseball cap, toting an AR-15 outside a Mexican ranch several years after his first prison escape in 2001.

The photo found in a BlackBerry seized at Chapo's mansion in Cabo San Lucas after his escape from Mexican and DEA authorities in February 2012.

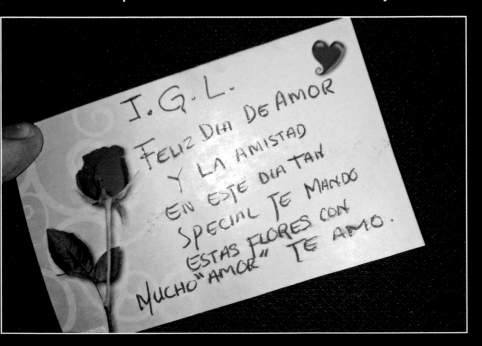

The card—headed by his initials, J.G.L.—that Chapo sent with flowers to his numerous girlfriends in Culiacán on El Día del Amor y la Amistad (Valentine's Day).

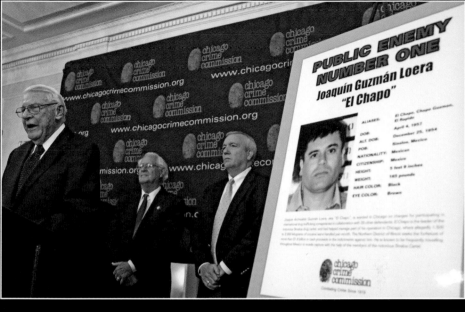

The Chicago Crime Commission names El Chapo Public Enemy Number One, replacing Al Capone, in February 2013.

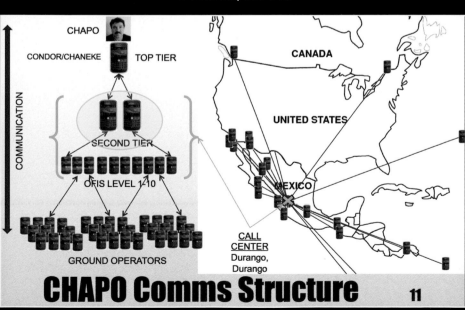

A diagram I created to show the communication structure of Chapo's mirror devices.

Video of Chapo interrogating a man tied to a post under a palapa.

Source unknown

An overhead shot of Duck Dynasty showing further construction: multiple palapas, a house, and a swimming pool.

Packages of cocaine stored in the tunnel at Safe House 3 along with fake plastic bananas used for shipping the drugs.

TOP: My Google map showing pertinent locations in Sinaloa, including
clandestine airstrips throughout the Sierra Madre mountain range, marked
with blue plane icons.

BOTTOM: My Google map showing the pings of the Top-Tier devices (yellow)
and other important targets and locations.

Imagery © 2017, Digital Globe; map data © 2017 Google, INEGI

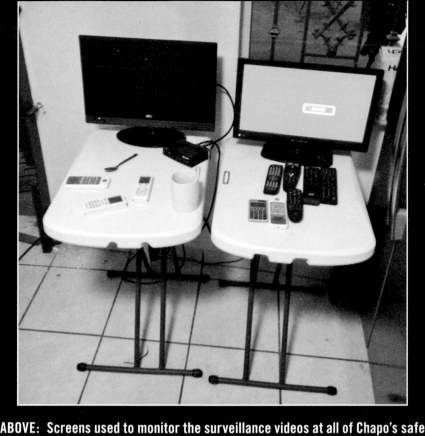

ABOVE: Screens used to monitor the surveillance videos at all of Chapo's safe houses—located in the garage of Safe House 1.

BELOW: Lying down for the first time on my potato sack cot in the makeshift SEMAR "barracks" in Culiacán.

SEMAR arriving at Safe House 2, making entry in the early morning of February 17, 2014. Taken on my iPhone.

Brady and me sitting in Chapo's driveway outside Safe House 3, taking a quick rest before the next raid.

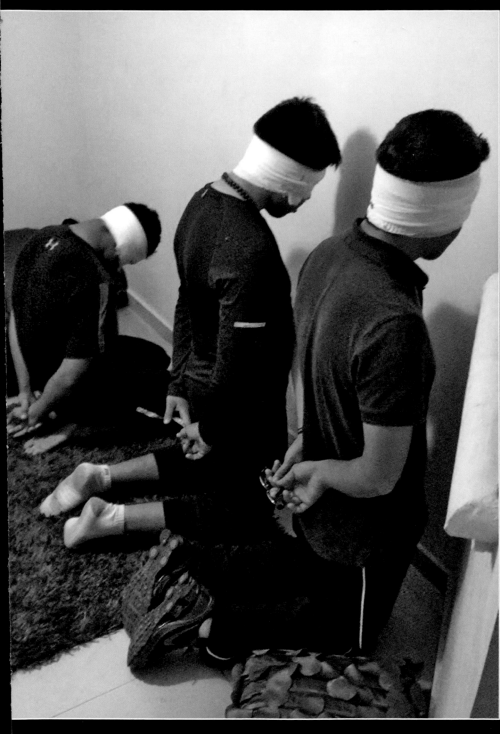

Several heavily armed subjects were detained inside Picudo's house.

Below Safe House 3, in a tunnel lit by fluorescent lamps, large quantities of cocaine were stored on makeshift racks.

Brady exiting the tunnel underneath the bathtub in Safe House 3.

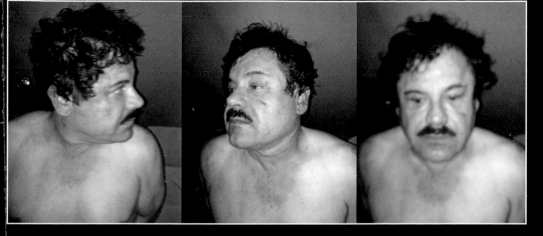

ABOVE: Three photos of Chapo taken on my iPhone inside my armored vehicle in the underground parking garage of the Hotel Miramar on February 22, 2014.

BELOW: Brady and me moments after the capture at the Hotel Miramar; I'm wearing Chapo's black ball cap and carrying the AR-15 rifle found in the hotel room with Guzmán.

Brady and me with Chapo in custody: the world's most-wanted drug lord during interrogation at the SEMAR base in Mazatlán.

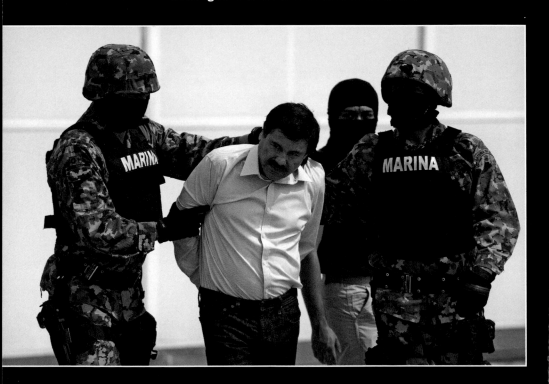

Chapo paraded in front of the world's press after he arrived at the Mexico City International Airport from Mazatlán on February 22, 2014.

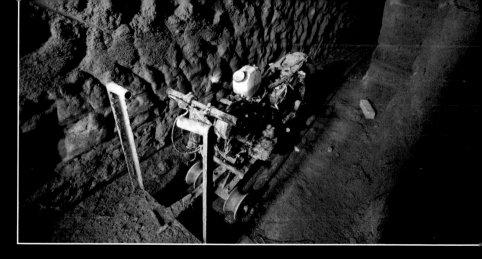

The 1.5-kilometer-long tunnel in which Chapo escaped from Altiplano prison on July 11, 2015. PVC pipe pumped fresh air throughout the passageway, and metal tracks had been laid so that Chapo could get away on a railcar rigged to the frame of a modified motorcycle.

AP Photo/Eduardo Verdugo

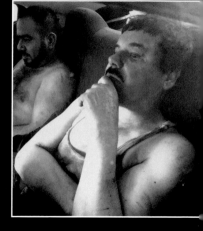

Chapo and Cholo Iván sitting in the backseat of a vehicle after their capture on January 8, 2016, in Los Mochis, Sinaloa.

Source unknown

Chapo sitting inside Cefereso No. 9 prison in Ciudad Juárez, Mexico.

Official Twitter account of Secretary of the Interior Miguel Ángel Osorio Chong (Mexico)

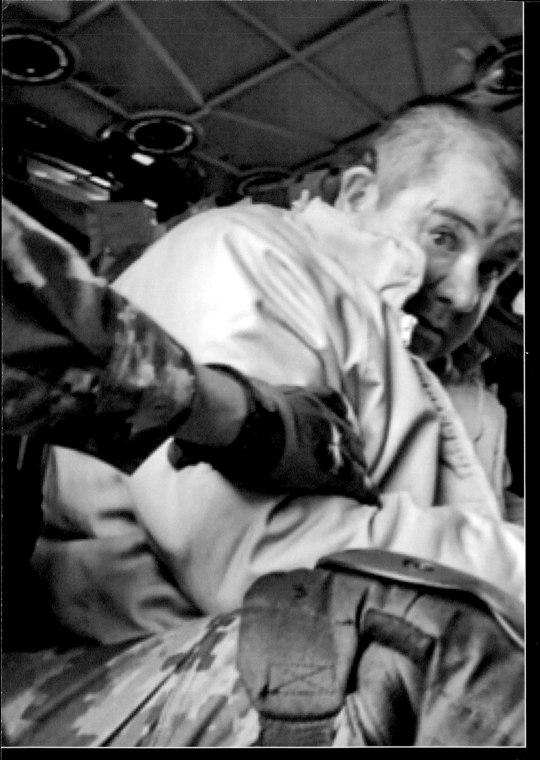

Chapo being extradited from Mexico to the United States on January 19, 2017.

Source unknown

Brady and I couldn't track the boys any further, but we could drill down on Chapo's language. What the hell did that name "Pichis" mean?

DECEMBER 24, 2013—10:34 P.M. I had just poured a cup of homemade eggnog, helping my wife finish wrapping up the last Christmas gifts for the boys, when my phone hummed with a text from Brady in El Paso.

"Got another Pichis run."

This time it wasn't just Güero and Ratón, but Tocallo as well. Chapo had instructed the three boys to meet his driver at the gas station in "Celis," and he'd take them the rest of the way in. My wife rolled her eyes as I turned toward the computer and began pinging.

"Celis" was the small town of Sanchez Celis.

"I can see a gas station at the south end, but all the pings dry up right around there," I told Brady.

It was one o'clock on Christmas morning, and Brady and I had been swooping around on Google Earth, searching that desolate area of Sinaloan farmland that faded into swamp further toward the Pacific, looking for any sign of Chapo's top-secret location.

Then it all began to fall into place. "He was ordering those airboats a couple months ago, right?" Brady said.

"Yeah, then he talked about building a pool near some palapas at Pichis." I replied. "So it's going to be someplace on the water— near the Ensenada de Pabellones."

"Found it!" Brady interrupted. "*Pichis.* An abbreviation. Short for Pichiguila duck-hunting club." Brady had already pulled up the club's website. I did the same, and came upon a home page advertising "the best duck-hunting in North America." The website even noted that President Dwight D. Eisenhower had hunted in the area, though before the club was established.

I was struck by the eerie throwback to my Kansas past. Chapo was a *duck* hunter? Guzmán hardly fit the profile of an outdoorsman, even though he'd been raised in the remote mountains of the Sierra Madre.

"Can you see him standing out in the middle of a swamp in knee-high waders, waiting for a flock of pintails to fly by?"

"Chapo's got no time for that shit. *Ducks?* Hell, no. He'd rather be hunting young girls," said Brady.

The club was located on the north end of the Ensenada de Pabellones. But my pings near Sanchez Celis were too far away from the club; Chapo couldn't be hanging out at the lodge. Brady and I continued our virtual flyovers, zooming in, zooming out, searching for any kind of man-made structure. Just after 3:30 a.m., I found myself hovering tight over two brown grainy circles resembling old palapas.

Jackpot!

I remembered my father teaching me a key lesson back when I was ten: whenever you hunt ducks, you want to sit on the "X." The X is the hot spot where the ducks are known to feed or relax. It was no different for Chapo.

Guzmán needed to leave his claustrophobic Culiacán safe houses to eat some carne asada, relax, sit under the stars in the middle of nowhere, and breathe fresh air for a few hours. A secluded place where he could meet with his sons to discuss cartel business face-to-face.

"Just found our X," I said. "It's the ultimate hit location." Brady laughed when I told him the name I'd given Chapo's secret lagoon hideaway.

"Duck Dynasty."

LOS HOYOS

THE KIKI ROOM WAS filled with suits. I was sitting at the center of the table, across from the head of the DEA. Administrator Michele Leonhart had flown down to the embassy to be briefed specifically on all the high-profile cases Mexico City agents had been working.

Leonhart was no stranger to stories of agents chasing down elusive kingpins. She had begun her DEA career as a street agent in Southern California and had worked her way up through the ranks before being appointed DEA administrator by President Obama in February 2010. The last time I had seen her was in 2006, when I walked across the stage at my graduation in Quantico. Before I began to brief her on the latest developments with Chapo, her words from my graduation ceremony once again rang in my head:

Go out there and make big cases.

I smiled, knowing that now, six years later, I was on the verge of fulfilling the promise I'd made with that handshake.

I filled Leonhart in on the tremendous relationship I had built with HSI and how far we'd come, showing her my Google Map and the heavy concentration of Top-Tier pings in the Colonia Libertad neighborhood.

"We have him down to a block radius, ma'am," I said. "And every week—or every week and a half—he comes out of Culiacán."

"Do you know where he's going?"

"Yes, to the coast," I said, "to a private hideaway he's constructing not far from the city."

Leonhart nodded.

"We're finalizing all the details," I told her. "I want to have enough intelligence to know where he's going to run to—should he escape capture again. We're almost there, ma'am. We plan to bring in SEMAR soon."

Following the meeting, I ran into the Regional Director Tom McAllister in the hallway.

"You spoke pretty confidently in there," Tom said. "Made some big promises."

"It's the truth," I said. "I just told it like it is."

I wasn't being cocky. I simply had no time to second-guess myself.

I had put in years of hard study back in Phoenix with Diego, and now I'd done all these grueling months in the trenches with Brady, Joe, and Neil, cracking the code of Chapo's mirrored telecommunication system. I could honestly say—without arrogance or bravado—that no one in history had ever been in a better position to capture Chapo Guzmán.

I now had the current location of the world's most wanted criminal. I'd saturated myself in all the tiniest details of Guzmán's life. I was sunk so deep into Chapo's world that I knew virtually almost every movement he made, and close to every order he gave throughout the day.

——

READING THE DAILY INTERCEPTS painstakingly, line by line, I could see a parallel obsession to mine: Chapo kept his eyes on *everything*, too, every single transaction and decision in his drug-trafficking sphere. In his own way, the kingpin was driven by a constant thirst for knowledge, just as I was. Chapo needed to understand every detail of his operation, even the most seemingly routine and boring things. This need for total knowledge bordered on a compulsion. Chapo was a boss who required complete control.

I needed control, too. Many evenings—nights when I should have already been home with my family—I found myself scrolling down through the summary of translated line sheets on my Mac-Book, hoping to find something—anything—I'd missed:

Chapo and Turbo, his Mazatlán-based maritime operator, discuss payments for a high-speed cigarette boat that needs to be sent to San Diego for several months to have new motors installed. Turbo asks Chapo to deposit the money into his wife's account and says that he still had not received his $10,000 bi-weekly payment for his expenses. Turbo also tells Chapo that he's looking at acquiring buoys equipped with GPS—clearly as a result of the recent 622-kg cocaine loss at sea.

Maestro, a pilot based in Chiapas, asks Chapo to send a payment for some unnamed military commanders. Chapo replies that he'll be sending a total of $130,000 in the plane: $40,000 to pay each commander for each event and $10,000 for the jet fuel.

Raul—an operator in Panama—notifies Chapo that he's found a place on a hill five kilometers from the border with a ranch and room for a clandestine airstrip, but he will need heavy machinery to clear it.

Raton is requesting "20 rolls"—$200,000—for logistics in moving 20 tons of marijuana in a tractor trailer towards the U.S. border. Chapo tells him that Picudo will deliver it.

Chapo reminds Raton that they shouldn't plan on having their BBQ outside Culiacán because there's been a lot of military and police activity outside the city. Chapo suggests having it at his son Güero's house instead and they can order Chinese food.

Chapo instructs Ciento, one of his gofers in Culiacán, to go get the press from Pinguino's ranch so they can make the squares—kilos—and to be careful because military and law enforcement units have been spotted in the area of the ranches.

Chapo asks his accountant Oscar how much cash he has on hand. Oscar says $1,233,940 not including what Güero recently gave him. Chapo instructs Oscar to make the following transactions. To give $200,000 to Ciento and have Oscar send him a confirmation after the transaction at the bank. Then he asks Oscar to give Pinto, still another Culiacan worker, 4,190 pesos to repair a car.

Chapo tells his son Tocallo that he'll meet with him tomorrow.

Chapo asks Araña how many loads were taken from La Cienéga, an airstrip outside Culiacán. Araña runs down through a list and reports that there were ten trips taken in the last week at $2,000 each.

Chapo tells Kava to hurry up with the deeds for the nine properties they'd been given. Kava tells him he's headed to Tijuana to conduct an evaluation of a site, most likely the start of a new super-tunnel. Then he discusses with Chapo the purchase of a casa de cambio—money-exchange—that's going out of business. Chapo's interested. Kava's also going to look at some other properties in Mexicali and San Luis.

Chapo speaks with an advisor named Flaco who's updating him on a hearing in the port city of Lázaro Cárdenas regarding one of the cartel's boats that's been seized.

Chapo sends flowers and a local five-piece band called Los Alegres del Barranco over to a 28-year-old girl in his neighborhood to serenade her on her birthday.

AND, LIKE CLOCKWORK, every morning Chapo would receive an intelligence report from one of his most trusted advisers; Brady and I knew him only as "Sergio."

Lic-F may have been Chapo's eyes and ears in many aspects of the DTO, but we recognized that Sergio was a critical player regarding Chapo's personal safety. For months, Brady and I would watch as Sergio reported the detailed movements of Mexican government operations—military and law enforcement—both inside and outside Sinaloa.

"Today the reconnaissance flights are scheduled to begin at 10:00 hours and run through 14:00 hours. One helicopter to the Cruz de Elota zone, another to Jesús María, and the final to the Navolato area."

"Four *sapos*"—toads, a coded reference to the green uniforms of SEDENA troops—"will be conducting patrols in the Cañadas, Las Quintas, Loma Linda, and Villa Ordaz neighborhoods today."

"Federal Police will be leaving the airport this morning and there will be movement from Mazatlán to Los Mochis—they're looking for meth labs."

"Two soldiers will escort a panel van carrying monitoring equipment through Culiacán."

The level of detail was so precise that it seemed to me as though Sergio was copying and pasting directly from the Mexican military's daily operational plans.

"Sergio's got people greased everywhere," I told Brady.

"Yep," Brady said. "Chapo's got advance notice on every single movement in Sinaloa."

Brady and I had almost become numb to the military leaks, after reading them every day for months. We hoped that Chapo had become numb as well.

"There's no way in hell he has time to stay on top of all of this intel," Brady said. "As busy as he is every day running the DTO."

"Regardless, he's got his safety net in place," I said. "He'll know immediately if there's even a whiff of an operation against him."

Brady interrupted our conversation with a freshly translated line sheet.

"Condor just told Naris that Chapo wants some sushi. He's to deliver it to Los Hoyos," Brady said.

"Pinging him now," I said.

"Condor won't quit," Brady said. "He's now sending the poor fucker back out for plastic spoons and a couple bags of ice."

When Chapo wasn't locked in his bedroom with his latest girl, he was all business, 24/7, running the day-to-day of his organization without even a break on Sunday.

"Naris is getting fed up," Brady said. "Check this out: 'I'm spending the day with my family,' Naris says. 'You can tell El Señor I'm not doing his bitch work today.'"

"'I'm not doing his bitch work,'" I repeated, laughing so hard that I forgot to send the coordinates to Brady.

When I did send them, Naris's ping came back with a successful hit.

"Same neighborhood, brother. Colonia Libertad. Right within that same block."

"What's Los Hoyos?" Brady asked. "'The Holes'—a street name? A stash house?"

Los Hoyos.

We both thought it over. Silence on the line. Both of us racking our brains . . . Then, once again, the word puzzle slotted into place. Brady and I said it in perfect unison:

"Tunnels."

This would certainly fit the profile of a man known as the tunnel king.

"Remember what I heard from one of my sources? Inside one of his safe houses, Chapo has a tunnel—entrance is underneath a bathtub."

"Yeah, of course, I remember—the tunnel underneath the tub."

"Bet anything that's why they're calling the place Los Hoyos."

Whatever Chapo meant by Los Hoyos, I kept pinging Top-Tier and Naris, zeroing in tighter on that dusty block in Colonia Libertad.

"My pings are close, man. His pattern of life is clear. Now all we need to do is find a door."

I KNEW JUST THE GUY I'd need to call once we were ready to go operational.

United States Marshal Leroy Johnson, from Mississippi, had been with the Marshals Service for years and had a reputation as the foremost expert in tracking the phones of fugitives. He worked primarily out of Tennessee, but he had traveled all over the United States and on international missions, too. Leroy had conducted enough counter-narcotics capture operations in Mexico that he'd picked up the Spanish nickname "El Roy." His southern accent was as thick as his build—he stood six feet and a solid two hundred pounds—and he shared my passion for a manhunt.

"This guy is crazy," I told Brady. "And fearless. He'll walk through the worst neighborhoods in Juárez to put cuffs on a bad guy."

I knew that if we were to conduct any capture operation, we'd need Leroy and his team of marshals on the ground.

By this point, the pressure was beginning to build, from Washington, DC, to Mexico City. Word had slowly trickled out among DEA and HSI that Brady and I had Chapo pinned down to less than a one-block radius. Upper brass from both sides were insisting on a final coordination meeting. I could feel the bureaucratic tilt—like at the embassy when it had shifted its weight during the earthquake.

I was now spending more time trying to appease my bosses and coordinate a complex interagency meeting than focusing on the plan to act on our intelligence. On top of that, the CIA's counter-narcotics team began showing up on the floor of my office, providing snippets of stale intelligence while asking probing questions, all in an attempt to collect anything they could about my plans.

ON JANUARY 16, 2014, at 6:53 p.m., I received a text from Brady.

"Chapo's headed back down to Duck Dynasty. Looks like he's already en route."

"Goddammit," I texted back. "We should have SEMAR on board by now. We're not going to get many more chances like this."

I knew that Chapo was at his most vulnerable leaving Los Hoyos behind in Culiacán, traveling light with Condor and maybe Picudo or some other bodyguard. I had been studying recent satellite images of Duck Dynasty, too. The photos revealed multiple palapas, a newly constructed pool—complete with a swim-up bar—and several other small outbuildings. A construction worker or two could still be seen on the property. This corroborated the line sheets: palapas, a new pool, the need for an airboat—all things Chapo had discussed randomly over the months were now making sense.

DUCK DYNASTY WAS in the middle of nowhere; the terrain was flat and easily accessed from all directions, so it was the perfect place to launch a capture operation.

But the clock was ticking. The possibility of a leak getting back to Chapo was too great. The only thing predictable about Chapo Guzmán was his unpredictability. He could switch up his pattern at any time and simply vanish up into the Sierra Madre for months.

I knew we couldn't wait any longer—we had the whole investigation wrapped up in a package with a big red bow on top.

"Think we can chance it?" Brady asked.

"Fuck it," I said. "Yes, let's bring in SEMAR right now. I'm canceling the meeting with the brass."

We had no other option: it was time to deliver.

MY FIRST CALL WAS to US Marshal Leroy Johnson, up in Tennessee.

"It's a go," I said.

"A capture op?"

"Get down here."

"On my way," Leroy said.

That next evening, I sat down with Leroy over a couple of bottles of Negra Modelo in a quiet bar across the street from the embassy, flipped open my MacBook, and showed him the heavy concentration of pin markers on my Google Map.

By now there were so many that you could hardly make out the city of Culiacán: it was completely covered with colored pins. I zoomed in, describing the meaning of each color and icon.

"Yellow pins are my Top-Tier pings," I explained. "Red pins are for the sons—also all of Chapo's other operators. Blue are points of interest—any important locations mentioned in the line sheets. Tower icons are cell phone towers."

"Good," Leroy said.

"An 'M' is for meeting places, or pickup locations frequented by his couriers. The red circles are where we believe he has other safe houses. The little planes mark clandestine airstrips—there are hundreds of them."

"And the yellow stars?" Leroy asked.

"Those are my closest pings—the ones with the tightest radius." I zoomed in tighter on the Colonia Libertad neighborhood. "Chapo is right here on this block."

"Unbelievable," Leroy said. "You've got his whole world mapped out, Drew. Safe houses, pickup locations, cell towers. Hell, I've never seen a pattern of life so dialed in. You've got him cornered."

"Not just yet," I said. "We've got plenty to work with, but we still don't have a door."

"Oh, we'll find it," Leroy said confidently. "Let's go get this son of a bitch."

EVERYTHING BRADY AND I had worked so hard to keep secret—even some of our own DEA and HSI people were in the dark about many of the details—was about to be exposed to the Mexican counterparts.

Admiral Raul Reyes Aragones, the top-ranking SEMAR commander in DF, commonly known by the nickname "La Furia" (the Fury, for the way he and his elite brigade would rip through and destroy Mexican drug-trafficking organizations), arrived at the US embassy in an armored Mercedes sedan, followed by his captain and several lieutenants.

Furia cut an impressive figure: early sixties, but extremely fit—he looked like he could still knock out fifty push-ups at the drop of a dime. Aragones's bald head was tanned and so shiny that it seemed to have been polished. He wore a crisp white-collared naval dress shirt, starched and without a single crease. His hands were soft and well manicured; when he smiled, his teeth gleamed white—*too* white, I thought, as if they'd been overbleached by his dentist. It was a salesman's smile.

"Quieres Chapo?" I asked immediately as the admiral leaned back in his leather swivel chair in the Kiki Room.

"Pues, claro que sí," the admiral said. "Of course we want Chapo! Tell me when and where."

I explained that Chapo and his entire organization feared SEMAR, whom they called *las rápidas*, "the fast ones." As I spoke, my chest felt suddenly tight, as if I'd just finished a six-mile run in that thin mountain air of the capital: I had so much to say, but I was almost too nervous to put my thoughts into words . . .

I'd spent years building up to this moment—how could I

suddenly let go? How could I begin to spill all the secrets and techniques it had taken me so long to perfect? No one had been told the *complete* picture—no one on earth knew everything that I knew—and now I was supposed to dump it all out in the lap of some spit-and-polish admiral I'd just met?

Realizing that if Chapo was going to be captured, we would need military help, I took a deep breath and stared the admiral in the eyes—this guy with the gleaming smile and immaculate uniform was the key to putting the operation into action.

"Chapo sometimes leaves the security of his stronghold in Culiacán for a lagoon getaway with a pool and swim-up bar."

"Why this location?" Admiral Furia asked.

"The proximity to Culiacán," I said, shrugging. "It's a short drive down—and it's remote from any prying eyes. Chapo likes to meet his sons—do business face-to-face. Some issues they just can't get done over the phone. It's an abandoned duck-hunting club called Pichiguila—Chapo refers to it simply as 'Pichis.' He's taken old palapas, had them renovated, and turned it into a pretty nice place. I call it Duck Dynasty."

There were a few snickers around the table.

"Dinastía de Pato," someone said.

"We've followed his sons—well, we've followed the *pings* on his sons' phones—and they take us right there near the palapas. So do the pings from the Top-Tier device."

"How do you know Guzmán is behind the phone you're pinging?"

"He's using *espejos*—mirrors—a layering system, which we've broken and infiltrated," I said. "Tiers, let's call them. In his terms, they're *secretarios*, and he's the manager—*el generente*. But one of two secretaries—Condor or Chaneke; we simply call them Top-Tier—is always by Chapo's side."

"Fine," Admiral Furia said curtly. "He comes down to this Pichiguila hunting club on the lagoon. Doesn't he travel with a lot of bodyguards? Our reports are that Guzmán's always got many heavily armed support with him—could be as many as one hundred—"

"That intel is *historical*," I interrupted. "It's out of date. At some point, yes, Chapo may have been traveling up in the mountains with that many armed bodyguards. Years ago, perhaps. Not now. He's moving fast and lean. He has a core group of loyal bodyguards—he's likely always armed, wearing a bulletproof vest, maybe in an armored vehicle. But no—those reports of hundreds of armed cartel men driving in a convoy of black Suburbans at all times? They are no longer valid. At this point, I can assure you that when Chapo comes down to those palapas he built near the Pichiguila Club, it's just with a couple of his most trusted men."

I could see that the admiral was hooked—and sure enough, by the end of my presentation he was ready to put all his men and resources at our disposal. We drafted an ambitious ops plan: simultaneous ground and air attack on Chapo's compound at the lagoon.

The element of surprise would be key: we would need to get SEMAR in perimeter positions in the middle of the night in order to capture the drug lord in a predawn raid. They would put four of SEMAR's top helicopters at a base in La Paz, near the southern end of the Baja Peninsula; they'd move in the elite brigades from Mexico City to bunk down at the SEMAR bases within Sinaloa, encircling Duck Dynasty.

"Once the helicopters and my men are in position, we'll wait until we receive the green light from you," the admiral said. "We won't move until you tell us that the Top-Tier device is there."

"Exactly," I said. "Once our pings indicate he's at Duck Dynasty, we'll let you know immediately."

"And your team of marshals will be joining us in La Paz, correct?"

"Yes, sir. They'll be with you in the event we need to chase the phone. If Chapo leaves unexpectedly, there's nobody better than El Roy and his guys to track him down. I'll also send Nico Gutierrez with you to be my DEA liaison from the ground."

SPECIAL AGENT NICOLÁS GUTIERREZ was a native Spanish speaker who sat next to me at the embassy and helped me ping phones and decipher some of the more unintelligible slang and misspellings in the line sheets. A former US Marine who was built like a defensive lineman, Nico was the perfect guy to be my eyes and ears in the field.

Gutierrez lived for capture ops like this. He already had his tactical gear packed and was ready to jump in with SEMAR.

Standing outside the Kiki Room, I ran into Regional Director McAllister again.

"Well, Drew, this is where your world begins to spin," Tom said with a smile.

"*Spin?* Sir, it already feels like it's about to unravel."

Tom was a seasoned DEA senior executive who'd led high-level cases throughout Latin America, Europe, and the Middle East. He understood better than anyone how hard I had worked to get even this far.

"I've filed a gag order with the chief of station here," Tom said. "CIA will not be allowed to talk about this operation with anyone."

I felt extremely grateful that Tom, and my other bosses, had allowed me to work without interruptions or any of the political drama that could often plague an investigation of this scope. They'd all been working diligently behind the scenes, ensuring that only those who needed to know were kept updated.

With the volume of intelligence coming in daily, Brady would need to remain in the El Paso war room, while I would establish a command center at the embassy with a group of intelligence analysts, along with all of my top brass. Brady and I both wanted to be on the ground with SEMAR, but we also knew that in order to keep the train from running off the tracks, we needed to be in our positions doing what we'd been doing nearly around the clock for nine months.

"SEMAR is ready to go. All Chapo needs to do is pop out of his rathole and come down for a meet one last time," I told Brady.

"From the moment he gets to Duck Dynasty, he's fucked," said Brady. "On that lagoon, he'll have nowhere to run."

THERE WAS NO turning back now—I'd told SEMAR everything they needed to know, and on January 19, the marines began making their first movements, flying into the base at La Paz and moving ground troops into the local Sinaloa bases in El Castillo, La Puente, and Chilango.

That same night, at 10 p.m., Brady called me.

"Goddammit," he said.

"What happened?"

"You've gotta see this. Just coming in."

It was a brand-new line sheet. Lic-F to Chapo.

"Sergio just met with the one from the water that has the special team in MEX. He's giving him ten rolls a month."

I felt a hollow ache hit my stomach. "Ones from the water"—code for the marines. And the "special team in Mexico City"—possibly Furia's brigade. Had our entire operation just been compromised? Ten rolls. That meant the contact was being paid $100,000 a month for intel. I tried to put my anger into words, but I couldn't.

"Hold on. It gets worse," Brady said.

By now I had pulled up the line sheets and translation on my own laptop.

Lic-F: Ahorita llegaron 3 rapidas del agua al castillo, puros encapuchados (son fuerzas especiales del agua) como que quieren operar en culiacan. Al rato nos avisa el comandante ya que platique con ellos a ver que logra saber.

"Three fast ones from the water arrived at El Castillo, all hooded ones (they're special forces from the water). Like they want to have an operation in Culiacan. The commander is going to let us know later once he talks to them."

Lic-F: Estan reportando 4 trillas grandes en la calma. Hay que estar pendientes pues no vayan ha querer cruzar el charco.

"They're reporting four helicopters in La Paz. We need to stay alert in case they want to cross El Charco."

"They're all over us," Brady said.

"'In case they want to cross El Charco,'" I repeated aloud.

El Charco—the pond. I knew that was code for the Sea of

Cortez, separating the Mexican mainland, including Sinaloa, from the long, thin peninsula where the La Paz base was located.

LEAKS. GODDAMN LEAKS. The steady drip, drip, drip had now turned into a deluge.

"Jesus," I said under my breath.

"He knows every move we're making," Brady said.

I reread the messages two more times, then stared at the multi-colored swirl of pin-markers on my Google Map.

The ones from the water.

Hooded ones.

Stay alert.

Crossing the pond.

Lic-F was reporting every military movement immediately to Chapo. I stared again at that long peninsula, at the La Paz base, then back at the heart of Culiacán. My pattern of pin markers blurred into a fiery kaleidoscope.

"Drew?"

The silence hung between us for a long time. Then I heard my voice, repeating, as if in a trance.

"Yes. He knows every move we're making."

PART III

LA PAZ

"DREW?"

I heard Brady, but I couldn't bring myself to answer.

Standing at the window, I could feel the heat rising in my chest—my neck and face were burning with frustration and anger.

I always tried to play the diplomat—"Switzerland," as Brady had said. I had been good at fostering smooth relationships and interagency collaboration; when there were flare-ups, it seemed like I was always talking people off the ledge. I never lost my cool. Getting angry never served the purpose of advancing the investigation.

But now I was on the verge of exploding.

I clenched the BlackBerry tighter in my palm.

"Someone in that room could be dirty," Brady said.

Chapo had a high-placed contact in Mexico City on the take for $100,000 per month? The level of corruption—the degree to which Guzmán had compromised the military and law enforcement brass, not just in Sinaloa but even in the nation's capital—suddenly seemed insurmountable.

The faces of the admiral, the captain, and the lieutenants quickly flashed through my mind. Just hours before, we'd laid it all out for them in the Kiki Room, and one of them could be corrupt?

Which one?

I was standing in the shadows of my living room in La Condesa, phone to my ear, looking out at the parked cars on the dark street below. As I stood there, peering out the window, seeing the ghostly trace of my face's reflection in the cold glass, I suddenly sensed that I was being watched.

Who was sitting in those parked cars below?

The new Charger.

The old Toyota.

The phone—was it even secure?

Someone could be listening to every word.

"What do you want to do?" Brady asked.

I took a long breath before answering.

"There's no other option," I said. "We have to confront the admiral—now. Get down here."

THE NEXT MORNING, I drove my armored Tahoe to the Mexico City International Airport and scooped Brady up curbside.

"We gonna pull everything back?" asked Brady. "Not work with SEMAR?"

"No. Without the marines—let's be honest—it can't be done. We gotta go see the admiral personally—at his shop."

"They're still warning Chapo on the lines," Brady said, and handed me his BlackBerry to read.

"We have to be very alert, like a pregnant bitch, dude," wrote Cholo Iván, the fierce sicario who ran the town of Los Mochis.

"We've never seen this many troops at this base—ever."

"Something strange is happening—be on alert."

"The fast ones, sir. The fast ones."

I DROVE DOWN to the south end of DF, swerving, gunning the Tahoe through the gaps in the congested traffic the entire way. At the SEMAR base, we passed through a single gate—manned by two young marines armed with automatic rifles—and were escorted upstairs to a large briefing room. As we approached the top of the stairs, I realized I hadn't slept or eaten in twenty-four hours. I was too unnerved by the thought of sitting down and facing the admiral; I still had no idea who the source of the leaks could be.

The conference room was surrounded on both sides by offices. The meeting was supposed to be private—just me, Brady, and Admiral Furia—but there were marines all over, coming and going from all the offices. Everyone was dressed in battle dress uniforms (BDUs), even the young marine serving us coffee and breakfast cookies, and a large projector screen displayed all the intel I had provided the day prior. Admiral Furia was sitting at one end of the long oak table, looking calm, in another one of his pristine white naval dress shirts. Brady and I sat at the corner next to him.

Brady and I exchanged concerned looks as additional officers entered the room. There were now twice as many officers sitting around the table as at the first meet at the US embassy. I didn't recognize any of these SEMAR brass. This was not a private meeting.

"Look at this shit," Brady whispered to me. "This place could be full of leaks . . ."

We were there to discuss a top-secret operation to capture the world's most wanted criminal, but the room was as bustling with activity as the flea market on Reforma that my wife and I took our sons to every Sunday morning.

"Yeah, too many fuckin' eyes and ears," I whispered back, but the admiral was impatiently gesturing for us to get on with it.

"We're here to discuss the compromise of intel," I began. "We've received messages that show the target"—I refrained from using Chapo's name aloud—"is seeing everything that's happening with your people up in Sinaloa and La Paz. Basically, he's getting real-time updates. Someone from an elite group of SEMAR in Mexico City is providing the information."

I showed Admiral Furia the messages.

The hooded ones.

The fast ones.

The helicopters.

In case they cross the pond.

Furia admitted that he knew SEMAR had leaks. He said it was not anyone from *his* shop, but he wasn't surprised that Chapo knew everything that was happening on the Pacific coast. He reassured me that he would do everything he could to find the source of the leaks immediately.

"Security is paramount to me," Furia said. "If this operation is going to be a success, we need the utmost in secrecy with this intelligence."

I had to stifle a smile. Looking around the room—the number of strange faces, the marine officers coming and going—made that statement laughable. I knew that if Duck Dynasty were compromised, and if Chapo aborted his plans to go to the Ensenada de Pabellones, we'd have no choice but to try to make entry into Culiacán.

A capture-op in Culiacán—just saying it aloud summoned images of a bloodbath. No one wanted that. The fatal firefight with Macho Prieto and his gunmen was too fresh in everyone's mind. But at this point, there was no turning back—I'd have to let the admiral know.

"What more do you have on the target's plans for the trip? Do you know where he'll be coming from?" the admiral asked, sipping his café con leche.

"At this point, yes," I said. "We have his location dialed in very tight. I've got him down to a block radius within Culiacán."

"Culiacán? You know where he is at this moment?"

"Well, yes—I don't have a street address, but I do know the neighborhood. We're sure it's the location of one of his primary safe houses."

The admiral exploded, shouting that we were holding out on him.

"No me cambias mi pañal!" "Don't change my diaper! And don't bottle-feed me!" Furia slammed his well-manicured hand on the oak table. "Trust needs to be established immediately."

I explained that it was not an issue of *trust*. I wanted to give SEMAR the very best intel we had. I played the diplomat now, apologized, and told the admiral that I'd provide all of our intelligence to him, exposing the network of safe houses in Culiacán where Chapo seemed to spend the majority of his time.

Admiral Furia took a deep breath, listening intently. Brady added that we meant no offense and were not being accusatory.

"Mira," Brady said, "we know we have dirty people working in our agencies, too. Even in the United States."

"No one wants Chapo more than *us*," Admiral Furia said. "I want to capture him more than *anyone* in this room. You Americans may not understand this, but his capture is more important to Mexico than the United States. He's a stain on our entire country."

I was impressed by the sincerity of his emotional outburst. The atmosphere in the room eventually calmed down. The leaks had been addressed, and Brady and I did exactly what the admiral wanted: *"abra las cartas"* (open the books). We disclosed years' worth of intelligence, taking the admiral meticulously through every detail of Chapo's secret world.

Just before we left the room, I grabbed the admiral's attention one last time.

"Señor," I said. "There's only one thing that can fuck up this entire operation."

"And what's that?"

"Los primos," I said.

The cousins—a well-understood euphemism for the CIA. I knew that the admiral had a couple of SEMAR intel guys on CIA's payroll, and sometimes they'd provide DEA info straight back to the spooks. The CIA could claim the intel as original and act on it, without coordinating with anyone.

The admiral summoned two officers—a captain and a lieutenant—and told them, rather theatrically, as if for Brady's and my ears: "Nothing goes to *los primos*. You understand? That's a direct order."

DRIVING BACK TO the embassy from the meeting, I called Nico to check in.

"How's it going over there, man?"

"Todo bien, mijo," Nico said. "These guys are ready to rock. I'll

be flying in on the lead helo, and El Roy and a couple of his guys will be in the one behind me."

"Okay, you've seen the leaks, right? The admiral's going to do his best to find the guy passing the intel here, and he's come up with another plan."

While we waited for Chapo to move, it was crucial that SEMAR concoct a plausible counterintelligence story: there were just too many *halcones* (hawks)—Chapo had a vast network of lookouts spying for him in Sinaloa. And so, SEMAR began spreading the word that they were conducting extensive training missions with helicopters, ground crews, and extra brigades on the Pacific coast, so that the heavy new military presence wouldn't cause Chapo's people any more alarm.

SEMAR had also coordinated government aircraft in La Paz, to further pinpoint the Top-Tier BlackBerry device the moment Chapo decided to break free from his Culiacán refuge and head down to the duck-hunting lagoon.

BRADY FLEW STRAIGHT BACK to El Paso to work the HSI war room.

A full week passed without any movement from the *secretarios*.

"Anything?" Brady asked.

"No," I said. "Top-Tier hasn't left that goddamn block."

"I'd be going nuts. Beyond stir-crazy. Imagine not leaving your house—not seeing sunlight—for a straight week?"

But if anyone was used to staying holed up in some safe house, it was Guzmán. Chapo seemed to be content staying in one location for weeks on end. His daily drug operations were seemingly not affected by the movements of SEMAR.

Still another week passed.

"He's not coming out," I said. "Our time's running short with SEMAR. Nico told me they've been conducting 'training missions,' flying the helos around Cabo, but even that's getting old. The marines are anxious."

"Hold on—this is just coming in," Brady said, reading a newly translated intercept. "Chapo's sending Naris down to Duck Dynasty to watch for any marine activity on the roads."

After pinging him, I could see Naris poking around, doing his own detective work. Several hours later, he reported right back to Chapo's safe house. Brady and I learned that Naris had spoken to some middle-aged rancher who neighbored the Pichiguila Club; the neighbor said he and his sons could hear a daily buzzing overhead. But whenever they looked up, they'd see nothing. Heavy activity at the La Paz base was being reported. By now Chapo was certain that something big was up—poor Naris was posted at the side of the road, eyes glued to the sky, waiting for the hum, like some London air marshal during the Blitz.

"Chapo may know *all* the movements of the marines," I said. "The only thing we still have going for us is that it looks like he doesn't know *who's* being targeted."

"You're right," Brady said. "If he did, he'd be long gone by now."

"And so far, there's been no mention in the line sheets of 'gringos,' right?" I asked.

"None."

"PACK YOUR BAGS," I said. "We need to re-strategize and motivate the troops. Let's go meet up with Nico and El Roy."

"All right," Brady said, "I can get down in a couple days."

Brady's wife had just given birth to their son, and I knew it wasn't an ideal time for him to tell her he had to bounce from El Paso.

"Sorry," I said. "No, I mean *now* now—we gotta keep the momentum up. That brigade's been there too long. Everyone's getting fuckin' antsy." The marines had been on standby at the base for two full weeks, just cleaning their guns and checking their gear, when all they wanted to do was rip their teeth into Chapo and his organization.

"We go meet the admiral face-to-face—I'm heading out tonight. Wednesday to Friday. Three days—real quick—we re-strategize this thing. Just fly into Cabo San Lucas; I'll pick you up there and we'll roll to La Paz."

The name of the air force base brought a smile to my face. It was officially known as Base Aérea Militar No. 9, La Paz, Baja California Sur, but everyone simply called it La Paz—the Peace. I didn't know what lay in store, but I was sure that I wasn't likely to see a moment's peace for a while . . .

I SPED HOME TO La Condesa to say good-bye to my wife and boys.

That night, I sat on the edge of my son's bed, reassuring him about his birthday party the following weekend.

"You promise?"

"I promise, buddy. I won't miss it."

No way was I going to be gone a full week. I kissed my son's forehead and told him I'd be back in plenty of time for the party.

Yes, a short trip, I reassured my wife. Three days max. Neither

Brady nor I was bringing any tactical gear or guns. I threw a couple of shirts, a pair of jeans, and some underwear into my carry-on and bounded down the stairs and into my black Tahoe.

ALL BRADY AND I needed to do was devise a plan B with the SEMAR brass in case Chapo never set foot outside of his safe house in Culiacán.

But there was too much unusual activity down in Sinaloa. Flight after flight left the BAM-9 La Paz base, circling Culiacán, attempting to zero in on the one-block diameter I had provided. Aerial images were fine, but we needed actionable intel on *addresses*—a few houses SEMAR could strike in lightning-fast raids.

Meanwhile, I was receiving current imagery of Duck Dynasty on my MacBook. From the photographs, I could see a hive of activity at the newly renovated cabins: workers were assembling a bridge out to a man-made island with a large palapa and working on a specially designed party house. The muddy swamp water that made the lagoon so perfect for duck hunting was obviously not something Chapo and his harem of girls would want to skinny-dip in.

THE CABO SAN LUCAS International Airport was mobbed with American tourists, flocks of college blondes in sarongs and flip-flops, itching to hit the beach. The guys were in baseball caps, surfer shorts, and wraparound shades, probably already nursing hangovers.

Hustling through the terminal, I thought back to the last time I was on a beach. Just six months earlier, I'd been watching my kids make sandcastles along the Florida coast, my toes sunk into the surf, when my BlackBerry, sitting on a striped towel, started vibrating—even in that peaceful family moment, Chapo couldn't help but intrude. I had decoded the messages right there on the beach, piecing together a murder plot from intel gleaned by DEA and HSI.

Chapo was getting ready to kill his own cousin, a forty-three-year-old named Luis, whom everyone called Lucho. Guzmán was slick; it wasn't going to be a public execution—no sicarios on motorcycles, toting AKs and wearing balaclavas. In fact, no one would ever be able to connect it back to Chapo.

Instead, Guzmán planned to simply send Lucho across the border into Honduras on a negotiating errand. Corrupt Honduran cops on Chapo's payroll would do a routine car stop—just as I had done so many times as a deputy sheriff in Kansas—but then they'd plant a hot gun and cocaine in Lucho's Toyota pickup, arrest him, and take him to a Honduran prison, where Chapo had arranged to have the guy shanked to death, making it look as though Lucho had been in the wrong place at the wrong time.

I had gotten up from my towel, walked down the beach, and, out of earshot of my wife and kids, called the DEA in Honduras. Sure enough, Lucho had just been arrested hours earlier. We got the guy put into an administrative segregation unit and managed to stop the murder plot. When I signed up to be a cop, I never thought I'd be saving the lives of high-ranking Sinaloa Cartel members.

Racing through the tourists in the Cabo airport, I was picked up by Nico in an armored Suburban tailed by a small convoy of *rápidas*—SEMAR's customized pickup trucks with machine guns mounted in the bed—and we headed up to the BAM-9 base in La Paz. I threw my bag on an open bunk, then scouted around for a few minutes, setting up a mobile communications center in a closet-size side room of the barracks, a room I dubbed Nerd Central: MacBooks, various iPhones and BlackBerrys, cables, and chargers snaked everywhere.

Brady landed in Cabo from El Paso a few hours later; we picked him up and drove to La Paz. We immediately sat down and briefed the field commander of the SEMAR brigade, two-star Admiral Antonio Reyna Marquez, a.k.a. "Garra."

Garra reported directly back to Admiral Furia, who had remained in Mexico City. I wasn't sure how this admiral had picked up the nickname Garra—Spanish for "Talon"—but it fit him perfectly. He had a tight cut of bristly jet-black hair, a sun-creased forehead, a hawklike nose, and high cheekbones that hinted at ancient Aztec bloodlines. He was no-nonsense, calm, and direct. Garra immediately demanded to know if we could still count on Chapo heading south, at some point soon, from Culiacán to Duck Dynasty.

"I wish I could give you a definitive answer, sir," I said. "Our biggest problem continues to be the intel leaks. Any luck finding the source?"

"No," said Admiral Garra. "We're still looking into it."

WITHIN A DAY, Brady and I had fully bonded with the brigade. We were the only white guys in civilian clothes on the base—and glaringly stood out—but then Captain Julio Diaz came around the corner with boxes of marine-issued BDUs and sand-colored combat boots.

"You guys need to get out of those clothes and blend in," Captain Diaz said. "Too many eyes around here."

I remembered Diaz from the meeting—every one called him El Toro. I'd liked him from the moment we met back in Mexico City.

"El me cae bien," I told Brady. "He falls on me well." It was an expression I had picked up in Mexico that meant you got a good vibe from somebody. "Just look at him. El Toro's always so amped up, ready to rock—he can't even sit still."

The bunks were tight, but I was thankful SEMAR had stuck us in the officer barracks with air-conditioning and decent showers. Brady and I joined up with Leroy and Nico for dinner—everyone, including the eighty marines from DF, ate in the same crammed mess hall.

"We gotta get over there," said Leroy, taking a few slurps of a funky fish stew. "Fuck this waiting around. He ain't coming out."

"Shit," Nico said, laughing. "What's the rush? I get a couple runs in around the base every day. Take a helo ride and buzz some beaches with these guys. Life's good."

BACK AT THE BARRACKS, we found out we weren't alone.

Two young DEA agents had arrived and were unpacking their bags. These were the two case agents leading the investigations of Mayo Zambada and Rafael Caro Quintero (or "RCQ"). A judge had recently granted Caro Quintero early release from a Mexican prison after he had served twenty-eight years of a forty-year sentence for his involvement in the murder of Special Agent Kiki Camarena. He was now wanted again, and believed to be hiding north of Culiacán, high in the Sierra Madre.

Ever since I began coordinating with SEMAR for the Chapo op, there was talk in the embassy about sharing resources with other viable DEA investigations. It was official protocol—no way around it. Brady had been incensed when I'd prepped him a few weeks earlier about the possibility of sharing SEMAR with another investigation. "You gotta be joking," he said. "Share SEMAR? For what?"

"Other case agents are claiming to have actionable intel on their targets, too," I said. "Now that we've lined everything up with SEMAR, they're saying they'll be just as ready as we are to launch."

"Bullshit," Brady said, fuming.

"I know," I said. "But there's nothing I can do about it. The decision was made above me." I had struck a gentleman's agreement with the other case agents: whoever had the ripest intel at the time SEMAR was ready in La Paz, that agent had authorization to green-light a launch. Fortunately, I knew that the intelligence these arriving agents had on Mayo and RCQ wasn't remotely close to what Brady and I had put together over the past nine months on Chapo.

"We can only hope SEMAR doesn't get distracted with all these different targets on deck. We've got to keep them focused," I said.

I knew that Mayo frequented the mountainous area just east of Duck Dynasty; an attempt to capture him first would immediately kill any chance we had of Chapo leaving Culiacán. He'd never risk it, knowing that the marines were conducting raids just south of the city.

As they unpacked in the barracks, I approached the two agents.

"You both realize if we launch on Mayo or RCQ first, our Duck Dynasty op is done," I said. "Completely burnt."

They nodded, but it was clear it didn't matter to them. I could tell they were just thrilled to be in the middle of all the action on the base.

AFTER A COUPLE OF DAYS at La Paz, I knew there was no way Brady or I would be leaving anytime soon. Nerd Central was our new command center, and we had SEMAR at our fingertips in the event we needed to launch.

Nico drove Brady and me to the local Walmart, where we stocked up on groceries and a couple of extra pairs of underwear. A return date to DF was the furthest thing from my mind.

I called my wife from the Walmart parking lot.

"Baby, I can't leave," I said. "I don't know how long it will be—maybe a couple more days—I'll explain later."

"Okay, please be careful," she said.

"Love you."

My wife had been rock steady throughout my DEA career. If she was worried, she never told me. We were the perfect match that way: always accepting and embracing life's precious moments—even the scarier ones—as a team.

For the next few days, Brady and I were glued to the hardwood chairs inside Nerd Central, reading fresh stacks of line sheets and pinging Top-Tier into the late hours, after all the marines had gone to sleep. Most nights, we'd be the only ones left in the room, all lights off, working by the glow of our computer screens.

Lic-F continued to keep Chapo updated on SEMAR movements, only this time there was no delay—nearly zero lag time. Brady and I watched as four SEMAR helos—two Black Hawks and two Russian MI-17s—took off from BAM 9 to conduct another training mission.

"Here it is," Brady said, reading from his BlackBerry within seconds.

"Lic-F just told Chapo four helicopters left the base at La Paz, and it appears they'll be conducting a training mission near Cabo."

"Spot-on," I said. Talk about unnerving. I didn't feel safe on-base. It still felt like we were all under constant surveillance.

There had to be someone on the base who was dirty. But who?

IN THE MIDDLE OF our fifth night's dinner, I received a call from my group supervisor in Mexico City, requesting an update, though it quickly became evident what her agenda was. She was letting me know—not so subtly—that she was receiving pressure from officials within "G.O.M." (government of Mexico).

"Drew, G.O.M. is leaning on me," she said. "Where do you guys stand? We don't have *unlimited* time with this operation."

She said that SEMAR was immediately needed to go fight Los Caballeros Templarios—the Knights Templar Cartel—down in Michoacán. The Knights Templar were a violent threat—founded by Nazario Moreno, a narco commonly referred to as El Más Loco ("the Craziest One")—but the notion that they were a higher-priority target for the DEA or SEMAR than Chapo Guzmán was ludicrous.

"Knights Templar?" I said. "Sorry, but the G.O.M. is full of shit. We're talking about Chapo. I have safe houses within Culiacán; definitive locations. There's never been a higher-priority drug target in the history of Mexico. Since I've been on the ground here with SEMAR—with their *admiral*—no one has said a word about Michoac—"

"It's gotten—well, political," she said. "I can only push back so hard, Drew. SEMAR is talking about pulling all of their people out of La Paz."

I hung up and stared at Brady. "My GS," I said. "There's talk about shutting us down. Pulling SEMAR from La Paz."

"Get the fuck outta here," Brady said, polishing off his last spoonful of lentejas, a watery bean soup.

"I'm not joking. 'Getting political,' she says."

I took a deep breath. Typical—bosses say one thing and the ground troops report another. None of these rank-and-file marines had any intention of leaving La Paz to go after the Knights Templar.

Still, I felt like the entire operational plan was in danger of unraveling. The double-talk was bad enough; more potentially damaging was the number of politicians, bosses, and bureaucrats in both governments who knew of the pending operation. I pushed my bowl of soup away, did a head count, and quickly lost track: the administrator of the Drug Enforcement Administration, the director of Homeland Security Investigations, and down the ranks through their various underlings and supervisory agents in Washington, Arizona, California, Texas, and Illinois. And *los primos* knew, of course, most likely right up to the director of the Central Intelligence Agency himself.

Too many mouths, I thought. *And too many egos.*

I didn't know the precise source of the leaks, but the effect on our target was crystal clear. Hell, *Chapo* knew more of the precise up-to-the-moment details of what was happening on the ground— more of the exact marine operational movements—than anyone in the US government other than Brady and me.

"THE PLANE SPOTTED Mayo's courier," Leroy Johnson said excitedly. "They're following him. Southeast of Culiacán. I think we're going to launch."

El Roy knew that our intel on Chapo should have taken top priority. But Leroy simply couldn't take sitting around the barracks any longer. He'd watch movies in the bunk room until 4 a.m., and his mind always seemed to be racing. He was pumped up and ready to get out into the field with the marines and chase down some bad guys, regardless of who they were.

I turned to Brady. "They're saying it's a fifty-fifty shot that this courier could be delivering some food to Mayo in the hills. They're launching."

"Goddammit."

Brady and I both watched as Nico, Leroy, and his team donned the tactical gear hanging from their bunks and joined teams of marines boarding the four helos out on the tarmac. I followed them outside, feeling powerless just standing there, the ocean and the grit from the rotor chop smacking me in the face. There was no stopping this operation now.

I watched helicopters lift off and eventually disappear over the horizon of the Sea of Cortez.

LATE THE FOLLOWING DAY, the helos returned to base, and I watched as lines of exhausted marines filled the mess hall.

I'd already been briefed on the results.

They hadn't captured Mayo, but they did arrest a few of his men and seized a cache of AK-47s, M-16s, and shotguns found buried in fifty-five-gallon drums on a ranch near the initial target location.

"Not good," Brady told me. "Several office devices in Durango dropped already, and the lines are quiet."

Brady and I knew this was coming—the same thing had happened after the premature arrest of Alex Cifuentes. If all the mirrors dropped, we were on the verge of standing in the dark once again.

I saw the Mayo case agent packing up his bags in the barracks, nodding on his way out.

"Well, he looks perfectly happy," I said to Brady.

"All of that for a fifty-fifty chance? Good riddance. At least he's outta here, so SEMAR can focus on Chapo."

As we were walking back to the command center from our barracks, Brady stopped, staring at his phone. A brand-new message from Top-Tier—intercepted and translated in the HSI war room in El Paso—had arrived.

"Shit, we're still alive."

"What you got?"

"Chapo just told Naris to go out and buy some red satin sheets and take them over to one of the safe houses," Brady said.

Chapo was going about his daily business—spooked or not. And any fears he had about SEMAR's movements just south of Culiacán and in the skies hadn't put a dent in his love life, certainly not on Valentine's Day—known in Mexico as El Día del Amor y la Amistad, the "Day of Love and Friendship."

"He's ordered Naris to get dozens of roses for all his women and write the same message on all the cards," Brady said. "He even wants Naris to sign his initials for him: J.G.L."

"J.G.L.," I said. "Doesn't get any more personal than that."

There was no doubt in my mind now. Chapo was still inside that safe house, on that run-down block in Colonia Libertad.

There wasn't a moment's peace to be found in La Paz. I was still on edge after the recent turn of events with SEMAR going after Mayo.

Brady and I knew that we'd need to refocus SEMAR back onto the manhunt for Chapo, so we went directly to meet with Admiral Garra in his office. I could see that Garra was tired, his eyes dark and puffy; he was clearly disappointed by the results of the Mayo raid.

Garra seemed annoyed by our very presence in his doorway. He didn't say a word; he merely raised his eyebrows as an indication for us to get to the point.

"Señor," I said. "Top-Tier hasn't dropped."

"You've still got him within that block?"

"Yes, surprisingly, Top-Tier is still pinging in the same place," I said. "Chapo seems comfortable in Culiacán. We might turn this to our advantage. He must think that all the military activity in La Paz was for the mission launched against Mayo. He's still going about his business. He just had flowers sent to all his girls for El Día del Amor y la Amistad, but there's no way he's coming out. Not now."

"So you're suggesting . . . ?"

"Going to ground," I nodded.

"In Culiacán?"

The name of the cartel stronghold—often called the City of Crosses, for its makeshift shrines to hundreds of murdered narcos—hung between us for a long time in the command center barracks.

A mug shot parade flashed across my mind: Ernesto Fonseca Carrillo, Miguel Ángel Félix Gallardo, Rafael Caro Quintero, Héctor Luis "El Güero" Palma, Amado Carrillo Fuentes, Mayo Zambada, Manuel Salcido Uzeta (a.k.a. "Cochiloco"), the Arellano Félix brothers, Chapo Guzmán . . . virtually all of Mexico's most infamous narcotraffickers had called Culiacán home. Going into the capital of Sinaloa was a daunting thought, like trying to wrest control of Chicago away from Al Capone's grip back in the heyday of Prohibition.

I stared at Garra, and he at me. We both seemed to acknowledge that it was the only option, but we also both knew the immense dangers ahead.

Nothing like this had ever been *considered*, let alone attempted. For SEMAR, and for two American federal agents, leading the capture op in Culiacán would be like walking on the moon.

GARRA PICKED UP his phone and made a quick call to Admiral Furia in Mexico City. He turned to look at Brady and me.

"Pack your gear tonight," Garra said. "We'll leave at oh-eight-hundred hours tomorrow."

That evening, the marines threw together a quick going-away party in a remote sandy corner of the base, among the cardónes—giant cactuses—and blue fan palm trees. They lit a bonfire, and SEMAR had their own version of a food truck slinging plates full of carnitas, tacos de barbacoa, and the marines' favorite, tacos de sangre—soft-shelled corn tortillas filled with blood sausage.

Sitting around the fire, I thought back to when I was eighteen, those Thursday nights with our varsity football team in Pattonville, when we'd huddle around the campfire and share stories in preparation for the big game under the Friday night lights.

I sensed a similar camaraderie taking shape there at La Paz—jokes cracked in Spanish, blood tacos devoured, cold cans of Tecate downed one after another. All the marines were in high spirits, knowing that in the morning they'd be leaving La Paz behind for good.

I nodded at Brady. We were about to make the big leap.

We would be crossing El Charco and heading into the heart of Sinaloa itself.

THE NEXT MORNING, February 15, 2014, I woke before sunrise and lay on my bunk, staring at the ceiling. The more I thought about entering Sinaloa, the more I felt my gut tighten. I reached for my iPhone and texted my father:

"Can't even begin to explain what's happened the last week, Dad. We're going to have to root him out of his hole, and it's not going to be pretty. But it's our only option."

"When you going in?" my dad texted back.

"We're gearing up now. Moving bases and command center into enemy territory. We wave the green flag Monday," I wrote. "Going to burn the city down."

FOLLOW THE NOSE

I TOSSED MY BAG through the door of the DEA King Air and grabbed a seat on the left side of the aircraft; Brady, Nico, and Leroy followed close behind.

I could feel the momentum building; SEMAR had become reenergized. I watched out my window as the MI-17s loaded with marines began to lift off. But SEMAR wouldn't be following us— they were headed straight to the Batallón de Infanteria Marina No. 10 (BIM-10) military base, Topolobampo, Sinaloa.

In thirty-eight minutes the King Air crossed the Sea of Cortez and touched down at the Mazatlán International Airport, roughly 125 miles southwest of Culiacán.

I groaned when I saw our rig. Someone in the DEA Mazatlán office had lent us the shittiest armored Chevy in the entire fleet: a six-year-old Suburban with 200,000 miles on the odometer. Even the dark tinted film was peeling from the windows. I had specifically requested two armored vehicles, and this was what they gave us?

"The USG at its finest," I said, turning toward Brady, but there was no time to get stuck in anger or frustration. We shoved our bags into the back of the rig and jumped in.

"How're the lines?" I asked.

"Quiet," Brady said. "Too quiet."

"I wouldn't be surprised if he knows we're here already," I said.

Nico took the wheel and Leroy the passenger side, handing back to us a couple of beat-up M4s painted in desert camo.

"We may need these," El Roy said with a grin.

"About fucking time," Brady said. "I've been feeling naked since I crossed the border."

I flipped open my MacBook and pinged Top-Tier. No luck. I tried a few more times.

"I think it's off. Maybe dead."

"We're going to need a little more luck," Leroy said.

Brady called Joe and Neil back in El Paso and told them to begin digging for the next Top-Tier number.

"Hang on, boys," Nico said, slapping the dash. "I hope this old girl makes it."

We left north out of Mazatlán, shooting up the backbone of Sinaloa, eventually meeting up with two SEMAR *rápidas* along Mexican Federal Highway 15D, just south of Culiacán, who escorted us the remainder of the way to join SEMAR at BIM-10 in Topolobampo. The base was located on a small port along the Sea of Cortez, not far from Los Mochis—the stronghold of Cholo Iván.

The sun had already set, leaving a faint row of pink in the sky and casting a hazy glow across the highway.

Nico pulled over so we could take a leak. I got out to stretch my legs and found myself trying to read the expressions of the marines as they stood in the rear beds of the *rápidas*, dressed in full camouflage and body armor, carrying all their tactical gear and their black machine guns.

I suddenly realized that I had no idea which brigade these guys were with.

"Hope these guys are from DF," I said to Brady as we stood in the ditch, cars whizzing behind us on the highway.

"If they're local, yeah, we're compromised," Brady said. "He'll know we're pissing in his backyard."

Standing there in the open, I experienced another burst of paranoia: I imagined a couple of marines on Chapo's payroll walking up behind us, drawing pistols, and shooting us, execution style, right on the edge of that ditch.

"*Vámanos,*" I said.

The 245 kilometers should have taken us more than three hours, but Nico pushed the rig to ninety miles per hour. Along the way, we passed exits for Las Isabeles, Cinco y Medio, and Benito Juárez, suburbs of Culiacán that I'd studied for hours, zooming in on my Google Map.

The highway was eerily quiet now, pitch-black, its rutted blacktop strewn with gravel. I was finally on the same narrow road that Chapo and his sons drove to get to the secret hideaway on the Ensenada de Pabellones.

We were now just a fifteen-minute drive from the blocks that I had lasered into my memory—Chapo was at our fingertips . . . I could *sense* it now. A beacon, pulsating, emanating from the city's center . . .

We circled Culiacán, heavier with traffic—Nico swerving around a couple of tomato trucks, heading north. As we passed Guamuchilera Segunda, my phone and Brady's vibrated simultaneously.

It was HSI in El Paso—they'd broken through with a new number.

"Top-Tier is back," I said, smiling. "We're still in this!"

———

TWO HOURS LATER, just after midnight, we rolled into BIM-10.

The Topolobampo infantry base was perched high on a hill overlooking the dark waters of the Pacific. Out front, I read the marines' slogan on a large sign:

TODO POR LA PATRIA

"All for the Homeland." A sudden fog had rolled in, covering the military base in a thick white blanket. I could barely see twenty feet ahead of the Suburban's headlights.

I jumped out and took a deep breath of the foggy sea air—there was a different aura here at Topo than in La Paz.

I was hearing that old Metallica song in my head—like I would before every Tiger football game back in Pattonville, strapping down my shoulder pads, taking the field in those tense moments before kickoff. I didn't realize I was singing aloud—at a decent volume, too. I sang the verses of "Enter Sandman" as we hauled our bags through the fog and into the barracks, bounding up the stairs to the second floor two at a clip.

"The vibe here's different," I said.

"Yeah, I feel it, too," Brady said.

"These guys are ready to fight."

One of the baby-faced marines ran up and told us that Admiral Garra had called an emergency briefing in the command room for 1:00 a.m.

Brady and I were the last ones to arrive—SEMAR officers and other marines were already sitting around the conference table, and there was hardly space for us to squeeze in.

———

THE LIGHTS WENT OUT, and everyone was staring at my PowerPoint maps projected on the large screen.

Before I could say anything, a pair of SEMAR intel analysts took charge of the briefing. These were the same guys who I'd suspected were on the payroll of *los primos*. I glanced warily at Brady: I couldn't believe what I was hearing. The SEMAR intel analysts were trying to steer the operation back toward following up on the capture mission for Mayo Zambada.

"Mayo?" I said. "Again?"

"What the hell?" Brady whispered.

When I looked around the room, even in the darkness, I could detect nods—some of the SEMAR captains and lieutenants were buying into this bullshit. Even Nico and Leroy were standing on the other side of the room, going along with it all. I couldn't take it any longer—I interrupted one of the analysts.

"Hold up a minute," I said. "What are you guys talking about?"

"Easy, man," Brady said, taking me by the forearm.

I couldn't lower my volume.

"Mira! Listen to me: we have the world's most wanted fugitive *here*—at our fingertips." I stepped forward and pointed to the screen. "We've got Chapo dialed in to a block radius—and you're saying you want to switch up targets and go after *Mayo* again?"

I took a deep breath, remembering how badly we needed SEMAR's full cooperation, lowered my voice, and addressed Admiral Garra respectfully in Spanish.

"Señor," I said. "Our intel will never get better than this."

I wanted to say it even more bluntly: this could be the greatest counter-narcotics success in the history of Mexico and the United

States. We were only hours away from nailing the most wanted fugitive since the US Navy SEALs took out bin Laden.

"We're on the verge of something historic here, *señor*. In thirteen years, since he escaped from Puente Grande, no one has got closer to apprehending Chapo Guzmán than we are right now."

The room went silent.

My chest was heaving. I swallowed, glancing at the SEMAR intel analysts. I could hear Brady's heavy breathing, too, and—very faintly—the Pacific surf crashing on the cliffs just outside the barracks.

The admiral was weighing his options, eyes ping-ponging between the intel analysts and me.

After a long pause, Garra folded his hands decisively in front of him on the table. He'd made up his mind.

"Vamos," he said, calmly, *"a activar Operación Gárgola."*

Gárgola.

It was the first time I'd heard the word, which was Spanish for "gargoyle." Gárgola was the perfect code name for the capture op—*G* for Guzmán.

Duck Dynasty was dead; Operation Gárgola was in effect. The analysts had already sat back down and didn't say another word.

Someone hit the lights, causing everyone to squint. It was before 2 a.m., but no one was getting any shut-eye. Instead, the entire brigade rushed off to get to work.

The middle of the night was ideal for the raid: Nico would ride with a crew of marines in the Suburban, while El Roy and his equipment would roll with more marines in a black Nissan Armada. Nico's crew was essentially running security for El Roy as his rig cut grids through the Colonia Libertad neighborhood—the one-block radius—around the safe house where I was almost certain Chapo had spent the past twenty-four hours. Their sole objective was to find a door.

I was worried now about Cholo Iván. That killer couldn't wait to get his green light—he'd jump on any chance he got to pull the trigger. If Cholo Iván and his people in Los Mochis detected any movement from the Topo base down to Culiacán, south through his territory, things could rapidly spiral out of control. And if that happened, SEMAR, Nico, and Leroy would quickly find themselves in a massive gunfight.

I hugged Nico and Leroy. "Give 'em hell, guys," I said, just as I'd done so many times with Diego back in Phoenix before a major UC meet.

It was 3 a.m. as Brady and I turned to walk back into the command center through the fog.

We quickly set up shop like we'd done at La Paz, moving Nerd Central into Topolobampo. I pulled up a map on which I was tracking the phones of Nico and Leroy as they headed south down Mexican Federal Highway 1D, the orange icons on the Find My Friends app dotting along as they neared Culiacán.

"Nothing in the lines about Cholo Iván," Brady said. "Don't think that he knows our boys are rolling through."

"Good," I said, nervously pacing back and forth.

BUT AS SOON AS dawn broke, the city lit up with news flashes. Brady and I were following along with all the Top-Tier exchanges in real time. Chapo was getting updates roughly every twenty minutes from Lic-F and Sergio, who had their *halcones* on every corner, on every street, instantly reporting how many SEMAR *rápidas* were in and outside the city and precisely where they were patrolling.

SERGIO: Ahorita estan por la canasta bienen puro gafe de agua no traen intel andan en rg en 19 a ver k cae hay las teniamos monitoriadas duraron paradas en la col popular en la calle rio usumasintris y rio grijalba

"Right now they're in the basket [city]. They all come from the special forces of the water. They didn't bring intel with them. They are headed to the RG in 19 [Culiacán] to see what happens. We have been monitoring all their stops in La Colonia Popular on streets Rio Usumacinta and Rio Grijalva."

Admiral Garra had sent groups of *rápidas* down behind the Suburban and the Armada to run security, but they'd been instructed to stay along the city's edge, circling like distant sharks. They were to respond only if Nico, Leroy, and their crews were in trouble.

Sergio's message to Chapo continued.

Hay estan como escondidas toda la mañana y se movieron rumbo a la canasta

"They were hiding there all morning, and they moved toward the basket."

All the *halcones* in the city knew which cars and trucks didn't belong; but there was no way to do this stealthily in any case, no way to avoid having Nico's and Leroy's teams hunting within that block radius of Top-Tier to locate a specific door.

"Man, what's taking these guys so long?" Brady said, pacing near the rear of the MI-17 just outside the door of the command center.

"Don't know," I said, "but they need to hurry. The city's getting hot. They're not going to be able to stay in those streets much longer."

Nico, Leroy, and their crews had been circling La Colonia Libertad and the surrounding neighborhoods for more than nine hours already, but we were still no closer to locating Chapo's door than when we started. Pinpointing that Top-Tier device from the ground was more difficult than we expected.

Then an incoming message appeared from the HSI war room in El Paso on our WhatsApp group chat. It was from Chapo to his cook, who was using the code name "Lucia."

Lucia, aplasten la tina del bano. Y para ke tesalgas en el yeta con memo la aipa la tableta. La traes tambien

"Lucia, flatten the bathtub so you can leave in the Jetta with Memo. And the iPad tablet. Bring it also."

Lucia, bengase fijando ke no las siga ningun carro y borre los mensajes

"Lucia, when you come make sure no cars are following you and erase the messages."

I stared at Brady.

"'Flatten the bathtub?'" I said.

"He might've jumped in a tunnel already."

"Yeah, he's starting to panic," I said. "Our boys have got to be close—right on top of the house."

I called Nico to relay the news.

"Any luck, brother?" I said.

"No, man, it's been rough," Nico said. "Every time we get a strong signal we lose it. We've marked a few points of interest—but nothing solid yet."

THROUGHOUT THE AFTERNOON and into the evening, Chapo was getting increasingly detailed intel: SEMAR was intercepting local two-way radio traffic, and the city's *halcones* were calling out every turn the Suburban and the Armada made—right down to the exact color of the rigs and how many men in camouflage fatigues were inside.

"Lines are starting to drop," Brady said. "Second-Tier has gone down."

"Fuck, we're too hot, man."

Brady jumped back on the phone again and immediately switched up strategy with Joe and Neil in El Paso.

"We've gotta rove!"

Rove—a roving wire intercept—was the fastest way we could legally track the members of Chapo's DTO, who were dropping phones and turning on new ones while on the run.*

"Stand by," Neil told Brady. "We'll be back up in no time."

Joe, Neil, and their team in El Paso had been working nonstop, getting about as much sleep as Brady and I were, laying all the legal groundwork so they could get that roving intercept authorized quickly with the help of Camila, their lead prosecutor.

---

* Under United States law, a "roving wiretap" is a wiretap that follows the surveillance target rather than a specific communications device. If a target attempts to defeat surveillance by throwing away a phone and acquiring a new one, by moving, or by any other method, another surveillance order would usually need to be applied for. However, a "roving" wiretap follows the target and defeats the target's attempts at breaking the surveillance by changing his location or his communications technology.

Just after 9 p.m., my iPhone buzzed.

"Drew, the fuckin' cops won't leave us alone; they're all over us," Nico said. "They've tried stopping us multiple times. This entire city knows we're here. Everyone's tired and hungry—getting burned out. Dude, this shit isn't working."

Brady and I walked back out onto the helo pad. Brady lit a cigarette he'd bummed from one of the marines. This was our eighth time walking in circles around the MI-17, knowing that Nico and his crew badly needed actionable intel.

"Fuck it," I told Nico. "Naris is our next best option. Find him and he'll tell us exactly where Chapo is at."

"So go after Naris?" Nico asked.

"Yeah," I said. "Follow the Nose."

BACK INSIDE the marine command center, Admiral Garra was furious with us.

"What the fuck is going on? We're in the same damn position as when we started. Our guys are on the ground and haven't found *shit*. I'm getting major pressure from my people in DF, asking me what we're even doing in Culiacán. We can't go on much longer—a few more hours and I'm going to have to call it off."

I could understand the admiral's frustration; I felt it, too.

"Señor," I said in a quiet voice. "We have to go after Naris."

"Chapo's courier is our best shot, sir," Brady added.

"If we don't find Naris, then we can reevaluate," I said. "But if we grab him, I'm confident he'll tell us exactly where Chapo is."

Admiral Garra just stared at me and, without saying a word, left the command center.

———

NOW THE COMMAND CENTER was empty; all the other marines had walked out to get some sleep. It was just me and Brady alone, so I cracked open a bottle of Johnnie Walker Red. I'd bought it before leaving Mazatlán and smuggled it onto the base in my laptop bag, hoping it would be a celebratory bottle . . . I found some red plastic cups and passed one to Brady. My stomach ached for food. How long had it been since I'd eaten anything solid? Eight hours? Eighteen hours? I had no clue.

Brady and I had bloodshot eyes—neither of us had slept in two days. The SEMAR brigade was finally sacked out—a wall of exhaustion had hit them like a tsunami. I sipped the Scotch and glanced at the time and date on my phone: 12:00 a.m., February 17, 2014.

I cursed softly, shaking my head. A father's promise, broken: Before I left DF, we'd picked out the piñata, gift bags, and invitations for my son and his friends.

"Dude, what's up?" Brady said.

"Hold on—gotta text her," I said, exhaling. "It's the seventeenth." I thumb-typed as quickly as I could and hit send at 12:02 a.m.

Sorry I'm going to miss it baby. This week has been one of the hardest in my life. I'm a zombie, exhausted & missing you guys. Having one hell of a struggle here. Give my son a big kiss & hug for me and wish him a happy birthday. I love you guys so much.

I stared hard at Brady and we topped off our Scotches.
Nada más que decir.
Nothing more to say.

Brady had a newborn son he'd left his wife alone to deal with; I felt like shit for promising *my* son I'd be there for the birthday fiesta in DF . . .

I scrolled through the tunes on my iPhone. I felt like blasting "Enter Sandman," or anything by Metallica or Nirvana. Even some crazy narcocorrido by Los Tigres del Norte, something hard and thumping, cranked to max volume to numb the mixture of exhaustion and sadness.

But all the marines were knocked out—I could hear a couple of them snoring—so I opted for "Cool Jazz for Warm Nights" and took a huge slug of Johnnie Red.

Brady let out a laugh as the soft jazz sounds floated through the barracks; the place reeked of sweaty battle fatigues and musty boots. The 1957 song "Everything Happens to Me," by saxophonist Warne Marsh, was playing when a message came in from Joe, back in El Paso.

"We're back live," Brady said. "Attaboy, Joey! Keep 'em coming."

The first message in was from Chapo, now becoming increasingly impatient, asking Lic-F for a status report. He was rattled.

Los bolas del agua donde kedaron

no saves??

"The group of marines—where are they at now? Do you know?"

MY PHONE STARTED BUZZING. The time read 12:34 a.m.

It was Nico. No greeting—he sounded intense, out of breath. "What's Naris's number?"

I quickly rattled off the digits to Naris's BlackBerry, which I had committed to memory.

I suspected why Nico needed them—but asked anyway. "Why?"

Nico laughed. "I got the motherfucker right in front of me."

"Really? You got Naris?" I said, grinning, glancing at Brady.

Brady almost knocked over his plastic cup of Johnnie Walker.

"Yup," Nico said. "Dude with the big nose is standing right here—six feet away from me."

"Okay, where's he telling you Patas is at?" I asked.

Patas Cortas—Spanish for "Short Legs"—was our open-air code name for Chapo during the capture op.

"He's saying he's at the Three," Nico said.

Only moments earlier, Brady and I had intercepted a message from Chapo that we were in the process of deciphering:

Naris si cnl bas ten pranito ala birra y le llebes ala 5 y traes aguas.
Seme olbido el cuete ai esta en el 3 en la cautiba atras me lo traes

"Naris, go in the morning and pick up the birria and the keys and bring them to the Five. And bring some water. Don't forget the gun. It's there at the Three in the back of the [Chevy] Captiva. Bring the pistol to me."

I knew that Naris was lying; there was no way Chapo was still at Location 3—I knew for a fact that that safe house was now empty.

I told Nico what I'd just read:

"He's bullshitting you. Patas isn't at the fuckin' Three. No one is at Location Three. It's empty. He's at the Five."

I could hear Nico telling SEMAR that Naris was lying to them. Then he hung up. A couple of minutes later, Nico called back.

"Naris changed his tune," Nico said. "You're right. He's saying Patas is at the Five."

"Send every marine you have to the Five," I said.

LION'S DEN

LA PISCINA.

I remembered that earlier that day Chapo had sent Naris over to "El 5" to meet with his pool-cleaning guy.

I called Nico immediately.

"La Piscina," I said.

"What pool?"

"Patas has referred to the Five as La Piscina before. The house you're looking for has a swimming pool. I'm almost positive. I'll send you the coords of the area. Naris was just there this morning. Condor's pinging off the same tower right near there."

"Okay," Nico replied. "We're going to the Five. Naris is on board."

The door of the command center flew open.

"Vámanos!" Admiral Garra shouted, his eyes still tight, as if he'd just been woken from a deep sleep. *"Vámanos!"* he shouted again. *"Levantamos a Naris."*

I was pleased to know that Garra had been tracking the events through his own marines on the ground just like Brady and I had been getting updates from Nico.

Hearing the whine of the turbine on the MI-17 helicopter

outside was like a straight shot of adrenaline. Brady and I scrambled up a short flight of stairs to grab our few belongings, the remaining phones, and the laptop bag.

"Don't forget the vests!" Brady said, looking back from the narrow doorway.

From the cold tile floor, I scooped up the two old bulletproof vests we'd scrounged in Mazatlán. I tossed one to Brady, who caught it mid-stride.

Hopping down three stairs at a time, I exited the command center, taking a deep breath of salty ocean air. Running out onto the helo pad, I tried throwing on my vest but realized that I'd grabbed one that seemed designed for a small child. I couldn't get the straps loose, and I ripped the thing over my head in a frenzy.

With the MI-17 blades whooping a few feet above, I looked at Brady.

"This is it," I yelled over the noise of the chop. "He's fucked!"

We hardly looked like US federal agents now. Under the tan-and-black vests, both of us were again wearing the SEMAR-issued camouflage BDUs we'd put back on after we arrived at Topolobampo.

I made my way through the huge hole in the rear of the MI-17 and took a seat on the hard steel bench directly behind the right-side gunner. Brady took the seat next to me.

Admiral Garra's demeanor was calm—almost *too* calm, I thought. I'd been studying the admiral over the past weeks, trying to determine what made him tick. A seasoned SEMAR commander with decades of experience fighting Mexico's narco wars, Garra was like a grizzled bird of prey: always calm, even when it was time to pounce.

Without a note of excitement—as if we were just going to pick up some street-level dope dealer and not a billionaire kingpin who had evaded capture since 2001—Garra shouted over the roaring engine, "When we arrive, we'll put him in this helo and bring him back here for the interrogation."

"What we need right now are some guns," Brady yelled.

Yes—we both needed weapons.

I looked around the cabin to see if there were any extra rifles lying around. This was turning into a full-blown military operation, but Brady and I had fallen into Sinaloa so fast that we'd never had a chance to fully prepare. We'd given Nico and Leroy back the M4 carbines before they headed south into Culiacán.

Three full days now without sleep. The helicopter lifted off the pad, angling southbound along the Sinaloa coast, headed for the birthplace and stronghold of the world's most powerful drug cartel.

The night sky now glowed brighter than the cabin of the MI-17. The marines donned their tactical gear and loaded their weapons, including a Mark 19 grenade launcher hanging off the rear deck and two M134 miniguns punched through the ports on each side of the bird.

Then everything went strangely still. No one spoke.

The gunner nearest me held a green tactical light between his teeth as he checked his Facebook page on his cell phone. Here we were, about to capture the world's most wanted drug lord, and this guy was nonchalantly checking social media postings as if he were sitting on his couch back home. On my Black-Berry, I quickly updated my group supervisor back in Mexico. DEA management knew only that something big was heating up, but I hadn't shared all the details. Anytime US government personnel were embedded with a host nation's military, there was potential for a political firestorm; not every manager at the Drug

Enforcement Administration and the Department of Homeland Security Investigations was pleased that Brady and I had left La Paz and ventured over into Sinaloa.

In a few minutes, we'd be putting boots on the ground in Culiacán. Setting foot in this city was a life-and-death proposition for US federal agents and SEMAR forces alike. For a few moments I thought about what to write. The fewer details the bosses knew at this point, the better, so I opted for two words.

"En route."

I closed my eyes, my mind still racing as I tried to focus on the sound of the helo cutting through the ocean air.

Why hadn't we heard anything about what was happening on the ground?

Brady and I kept checking our BlackBerrys every few minutes, then we'd look to the admiral for an update.

Garra said nothing.

We were forty minutes into the ride, and still no one had heard a peep. If anyone was to be notified of a capture, it would surely be Garra. His SEMAR entry team would be first through the door and the first men to make contact with Chapo.

Maybe the marines already had him in custody and for safety reasons weren't advising command over the radio?

Brady's face, as always, was locked in his ruddy-cheeked scowl. Admiral Garra's remained inscrutable.

The MI-17 angled away from the coastline. I could make out the hazy glow of the city lights in the distance as we whipped past small homes and larger ranches.

"Quince minutos," one of the crew members shouted from the cockpit.

Fifteen minutes away. Just then, the red indicator light began to flash in the upper right corner of my BlackBerry. It was a message from Nico, with the SEMAR entry team at Location Five:

"Place is a fortress," Nico wrote. "Cameras everywhere."

The pilot began to cut figure eights over the city. I looked down at the streets below, deserted except for the frenzy of SEMAR's *rápida* pickup trucks, machine guns mounted in each bed and loaded with marines, crisscrossing the blocks as if they were beginning a search mission.

I could see a Black Hawk helicopter conducting a search over a different grid of the city, parallel to us. I looked to Brady and shook my head.

"Where the hell is he?" I said aloud, staring deep into a residential neighborhood below, as if I expected to see Patas Cortas jogging down the desolate street in his tracksuit. All I wanted was to get on the ground and start hunting. We were worthless in the air.

The old Russian aircraft banked hard right before making its quick descent into an empty city lot.

As we got closer to the ground, I suddenly began to lose my orientation—I hated the sensation. I'd always prided myself on my intuitive sense of direction. When I was in college, I often rode to the bar in the trunk of my friend's car because there were too many girls in the seats. Locked inside the dark trunk, I called out every single turn and every street name, until we got to the bar, never once losing my bearings.

But now I had no clue where we were touching down or what section of Culiacán we were in. Even with a full moon, I couldn't even tell north from south.

We jumped out the back of the helo just as it was landing. The marines exited rapidly and disappeared into the tall grass of the abandoned lot. Brady and I found ourselves alone now, unable to hear each other over the rotor chop, squinting as the dust and grit whipped around us. We quickly lost sight of all the marines; we even lost Garra. Brady and I had planned to stay glued to the admiral's shoulder throughout the capture operation.

I pulled my iPhone from my pocket and tried checking Google Maps to pinpoint our exact location. No luck. The Black Hawk was now attempting to land in the same tiny lot, and the blur of swirling dust made it nearly impossible to see.

"Jesus," Brady said, squinting. He'd spotted some of the marines jumping into *rápidas* off in the distance about two hundred yards away. "Let's go, Drew."

We began to sprint across the uneven lot, over chunks of broken concrete and weeds, as the Black Hawk landed. From the corner of my eye, I saw marines bursting out of the sides of the helicopter, covering their flanks with rifles drawn.

Brady was leading the way, with me right behind, humping my laptop bag.

I heard Brady yell, "This is a turning into a shit show."

"Yeah," I said. "No rifles. No radio. And if those trucks take off without us, we are fucked . . ."

WE WERE TWENTY YARDS away when the *rápidas* sped off. We began to chase them down, sprinting until we reached a gap in the chainlink fence that led to the street. But we were too late. The *rápidas* were long gone.

Out of breath, we rounded the corner and entered the empty street.

We could no longer see the taillights of the trucks, nor hear the roar of the helicopters behind us. All we could hear was our boots on the pavement and our rapid breathing.

I turned from side to side, trying to determine in which direction we were headed—but the pavement, trees, and buildings suddenly all blurred together in shades of brown and gray.

My gaze narrowed. I spotted a black silhouette about a hundred yards away.

I instinctively reached for my thigh holster to pull my pistol, but my hand just slapped my leg. No holster. No gun. The silhouette was getting closer. Was that the barrel of a rifle?

Then Brady shouted, "He's a friendly!"

As we got closer, I saw that it was a very young marine, posted up alone. He was slight, with a narrow pointed nose and brown eyes, his helmet far too big for his head. To me he looked like a twelve-year-old boy dressed up as a soldier. Brady and I jogged up to meet him.

"*A dónde van?*" Brady yelled. Where had all the *rápidas* gone?

The young marine shrugged. He seemed just as lost as we were, but we decided to follow him anyway. At least the kid was carrying a rifle. As we walked down the block, Brady asked in Spanish, "You have a radio?"

The young marine shook his head no. Brady was now staring hard at me.

We were exposed, and weaponless. There was no way to disguise our gringo faces; we didn't even have military hats or helmets.

And unlike in Mexico City, here in Culiacán no one could possibly mistake us for locals.

This is a fucking setup, I thought.

Chapo had paid off the military; we were stranded, without

guns or radios, in the heart of Culiacán, about to be kidnapped. The video of us being tortured and killed would be uploaded on YouTube before sunrise . . .

"Dude, we need to get back to the helo," I said. *"Now."*

But was the MI-17 even there? Brady and I abandoned the teenaged marine and started sprinting. We knew that if the helo took off we would be completely stranded, weaponless, and stuck in the middle of the lion's den.

SUDDENLY ANOTHER CONVOY of *rápidas* appeared, rounding the corner.

"Fuck it," Brady said. "Let's roll with them."

We ran toward the trucks, and a couple of the marines in the back were waving us on, so Brady and I jumped aboard. We squeezed into a double cab already holding six armed marines. I had no idea who these SEMAR guys were or where we were being taken. They were grittier than the ones I'd seen at the base in Topolobampo. I was pressed tight against a skinny kid with a dark complexion; he was smoking a cigarette, his helmet half-cocked to the side. Most of the others were wearing black balaclavas to hide their faces.

After driving for a few minutes, we parked in the center of a residential block, a typical middle-class neighborhood. As I jumped out of the truck, I looked up and down the intersection and could see more *rápidas* and marines posted on every corner. My nerves began to calm. Two marines handed Brady and me black balaclavas to cover our faces.

"Las cámaras," one of the marines explained.

The safe house had surveillance cameras inside and out, and everyone's faces needed to be covered before entry.

Brady and I walked up to a modern beige two-story house tucked between two other homes the same size. I was so disoriented, I still wasn't sure where our *rápida* had just dropped us off. Was this the block radius I'd been studying so hard on my Google Map? Were we even at Location Five? Neither Nico nor Leroy was anywhere in sight.

Brady and I walked cautiously through the open garage of the house, passing a black Mercedes four-door sedan, then stared at the badly damaged front door of the house. One of the panels of the door was missing, and the jamb was completely torn up in a twisted chunk of metal. The door had been reinforced with six inches of steel—SEMAR had clearly taken a long time to batter its way in.

I STEPPED THROUGH the entryway. The kitchen, directly in front, was furnished simply: white plastic table and folding chairs. Then I took the first immediate right through the living room and entered the ground-floor bedroom. Girls' clothes were scattered throughout the room. Lingerie, blouses, sweatpants, used towels, and open pill bottles littered the bed and the floor.

Brady and I slowly entered the adjacent bathroom.

It was dark and quiet, and far more humid than the rest of the house. I tried to flick the wall light switch; it was broken. Brady and I used the flashlights on our iPhones as we advanced.

There it was—unmistakable in the dim bluish glow.

"Dammit," I said.

Kava's handiwork.

"Look at this fuckin' thing," Brady said as we inched forward.

The large white bathtub, rigged on hydraulics, was propped up at a forty-five-degree angle. As we crept closer to the tub, an over-powering smell of mold filled my nostrils.

We gazed into a sophisticated man-made hole beneath the tub. A narrow vertical ladder led down to a tunnel that extended in the direction of the street, approximately ten feet under the house.

Brady descended the ladder first and made his way to the bottom. I was right on his heels.

The moldy air was so thick and hot that it was now hard to breathe through the face masks. Stooping, we both walked the length of the tunnel. It was extremely well constructed, rigged with fluorescent lighting and wood shoring. We continued until we reached a small steel door with an industrial-size circular handle.

Brady cranked the steel handle counterclockwise, revealing yet another dark tunnel. Trickles of sewage were running along the floor, and the five-foot ceiling caused us both to crouch down into a duck walk.

"Shit," Brady said softly.

We stared into the darkness.

We were looking at a gateway into the labyrinthine sewer system just below the city streets. Everything was pitch-black in both directions, except for a tiny pinhole of light a good twenty or thirty blocks away.

I tried to catch my breath. I looked in one direction and Brady looked in the other, hoping for any sign of life—a whispered voice, a cry, footsteps splashing through the fetid water . . .

Nothing.

"Gone," Brady said.

Chapo had escaped again.

THE DROP

CHAPO'S WORLD WAS UPSIDE-DOWN.

Now so was mine.

There was nothing to do but continue the hunt.

I scaled the tunnel ladder and slowly crawled out, ducking down to avoid hitting my head on the bottom of the bathtub.

Still no sign of Nico or Leroy.

I pulled out my iPhone and texted Nico.

"Where you at?"

"The Four," Nico replied. "Heading to the Three next. Meet us there."

I could see Nico's orange icon blinking about ten blocks to the east on the Find My Friends app.

"I bet Chapo could walk the sewer right down to the Four," Brady said.

"Yeah, he never has to see the light of day."

We walked into what had so recently been Chapo's bedroom and began rifling through everything—all the piles of clothes, towels, ledgers, miscellaneous notes, boxes of Cialis, Celebrex, and other prescription pills littering the room.

I only cared about one thing.

"Get me all the BlackBerry boxes and SIM cards you can find," I said. We needed anything that would offer clues about where Chapo had run and who he'd turned to for help in the final minutes.

"Jesus, they're everywhere," Brady said. There were more than twenty BlackBerry boxes in the bedroom alone. Brady and I quickly collected them into a pile on the bed.

I began snapping pictures of each distinct PIN number printed on the side of each box. As soon as I sent the PINs to Don, back in Virginia at DEA's Special Operations Division, they'd be able to get me the corresponding phone numbers almost immediately. Then I could get to work pinging the devices.

"There's a good chance Chapo's carrying at least one of these BlackBerrys," I said.

AS WE CONTINUED COMBING through the empty house, I ran into Admiral Garra.

"Ven conmigo," the admiral said abruptly, gesturing for Brady and me to follow him outside.

We jumped into another *rápida* and sped off in a small convoy, following all the other SEMAR trucks. Garra's face was determined, but his brow was starkly creased—it was clear he was still angry that Chapo had slithered away into the sewers.

It was only 4:30 a.m.—still too dark to see clearly when the *rápida* came to a stop—but as soon as my boots hit the gravel street, I knew precisely where I was standing. It was the exact block I'd been studying on my Google Map and high-res imagery for months.

The *rápidas* filled the street, marines piling out and swarming up the driveway. I stood back, taking in Colonia Libertad. I watched as a couple of marines led a man in a red-and-black polo shirt toward the house—even in the dim light I immediately recognized him as the courier Naris.

Naris was silent, head bowed, hands cuffed in front, leading the marines to a long brown steel gate, solidly built and electrified. I knew this was the same gate that Naris had been waiting outside for minutes on end, after running out to buy shrimp or sushi or plastic spoons for Chapo, standing there shouting, *"Abra la puerta!"* and pleading for Condor to let him inside.

This time it was Naris using a set of keys to open the gate for the marines. He was fully cooperating now. I stepped through the gate with the marines and turned around to look back at the street. I was more dialed in than I had realized: it wasn't a *one-block* radius—my pin marker was a mere twenty paces away, across the street from Chapo's driveway, close enough to hit the garage with the toss of a football.

The entry team smashed the reinforced steel lock of the side door, and dozens of marines flooded inside. I followed on their heels, stepping fully into Chapo's world now. This was his primary safe house, the one in which he had spent ninety percent of his time.

I stepped into the first bedroom on the right, scoping everything in the room, taking more photos of BlackBerry boxes and SIM cards. The marines were already beginning to turn the place upside down.

I heard Brady shouting:

"Why aren't they in the sewers? Get in the fuckin' tunnels!"

I knew there was no stopping it now, no way to tell this SEMAR machine what to do.

There was a bag of meth on the kitchen table. This was odd—snorting ice didn't seem like Chapo's thing. In the master bedroom, down the hall, I ran my hand across Chapo's long line of dress shirts and kicked at the more than fifty boxes of shoes stacked high in the closet. There were a couple of expensive watches—one was a rose-gold Jaeger-LeCoultre chronograph with sapphire crystal, brand-new in its box from Le Sentier, Switzerland.

Aside from the designer shoes and a few elegant Swiss timepieces, though, everything seemed to have been purchased in bulk at Walmart.

"Same cheap vinyl sofas," I said. "Same white plastic table. Same folding chairs."

I was surprised to see that Chapo afforded himself so few luxuries. This house was no better than Location Five. These were cookie-cutter homes, completely utilitarian and almost certainly designed and crafted by Kava and his crew.

I followed El Toro, the fiery SEMAR captain, along with a few other marines into the bathroom adjacent to Chapo's bedroom. El Toro was pushing Naris forward, clearly on a mission.

I rounded the corner and came face-to-face with Naris.

His prominent nose was now bright red.

Chapo's courier, hands still cuffed, moved over to the sink and stuck a small shiny object—it might have been a paper clip—into a hole near the electrical outlet next to the mirror. There was a crackling sound—for a moment I thought Naris had given himself an electric shock, but he'd somehow activated an internal switch, triggering the hydraulics.

The caulk liner around the bathtub began to crack away. Naris walked over and grabbed the top rim of the tub with his cuffed hands, giving an awkward lift until the power of the hydraulics took over. The stench of mold and sewage once again filled the bathroom as the entire tub was raised up to the same forty-five-degree angle I'd seen back at Location Five.

A SEMAR lieutenant—everyone called him "Zorro"—kept barking at his troops. "*Mira!* Strip your gear, get into the tunnel, and *find* the motherfucker!" No time to souvenir-hunt, Zorro said. This was their chance to catch the world's biggest drug lord.

Zorro was the first down the open bathtub, descending into the nasty sludge-filled sewers. He quickly disappeared with his team. But I knew that Chapo was in the wind. The guy was as slippery as a sewer rat. He'd likely emerged from some drainage hole more than an hour earlier.

I WALKED OUTSIDE and saw a large blue tarp above my head, a makeshift canopy that spanned across the driveway, from the guesthouse to the roof of the main house. Chapo clearly knew that there were always eyes in the sky watching him.

The guesthouse—equipped with a bathroom and a queen bed—had been built near the far back corner of the small lot, no more than thirty feet from the side door of the main house. By the time I poked my head in, SEMAR had ripped the place to shreds. I figured that this might have been the residence of Chapo's full-time cook or maid—everyone in Mexico who could afford it seemed to have a live-in housekeeper—but it could also have been where Condor stayed during his fifteen- to thirty-day shifts as *secretario*.

After an exhaustive search of the entire property, the swarm of marines shifted gears without warning, filling the streets and jumping back into their *rápidas*. I grabbed Brady by the shoulder as we were leaving and pointed to the white Chevrolet Captiva sitting in the driveway—we had almost overlooked it. It was the same Captiva that Chapo had ordered Naris to get into a few hours earlier, with instructions to bring him his pistol.

NOW WE WERE SPEEDING over to Location Two—it was only a few blocks away, so close we could have walked there. When we arrived, I opened the truck door to exit and paused. I was about to stand on another piece of asphalt I'd been studying for so long in my satellite imagery.

Until that moment, everything had been unfolding at lightning pace, but the initial shock and adrenaline rush were now wearing off. I began to realize how vulnerable we were. At any instant, we could all be ambushed, taken on in a hail of gunfire right there in the street. I pictured streams of Chapo's enforcers and their gunmen, other traffickers, dirty local cops—anyone with weapons—rounding the corner and opening fire. There would be nowhere to run.

I looked around in the truck for an extra rifle, pistol, or even a knife—nothing. A wave of fear rolled through me. I jumped out of the *rápida* and hustled and worked my way in among the mob of marines headed for the door—I figured that it was safest to be tucked in among the troops.

"The Two," as Chapo referred to it, was constructed like Safe House 3—with heavy concrete walls rising high to deter observation from the street and black wrought-iron fence completing the gaps—yet was similar enough to the rest of the neighborhood homes not to stand out. "The Two" was painted white, with a couple of large palms stuck in the walkway just inside the door, along with an attached single-car garage. I was familiar with this location, too. I'd studied the detailed overhead photographs of this very place back in La Paz when I was zeroing in on the pings of Picudo.

Once inside, we found the house almost bare. The three bedrooms each had a bed, but there was little or no other furniture.

"This place looks like a straight stash-or flophouse," I told Brady.

SEMAR had found yet another identical tub on hydraulics in the bathroom off the main bedroom, only this one was nearly impossible to enter. It was stuffed with more than one thousand individual football-size packages wrapped in brown tape and marked with a four-digit number that appeared to be the weight. Methamphetamine. In the end, we would calculate that more than three tons of meth were jammed into the tunnel.

It made no sense to me. Tens of millions of dollars' worth of meth just sitting there getting moldy in the bowels of Culiacán?

"Maybe Chapo's cash-out stash?" I asked Brady.

"Could be," Brady said. "Given the street value of that stuff, he could live on the run for years . . ."

The sun was coming up fast, the Culiacán horizon growing brighter by the moment, the streets starting to come alive. I stood for a while out in the street, with a few marines holding outer security. I noticed a grade school directly across the way; soon it would be bustling with children.

Something else, too: there was blaring music wafting down the hill. Who would play banda so loud at this hour? Was it a signal of some sort from Chapo's loyalists? A call to arms?

BEFORE I KNEW IT, Brady and I were back in the convoy, riding with Nico and Zorro in the armored Suburban, speeding north to yet another location.

It was the first chance I'd had to talk to Nico face-to-face since the frenzy of the first predawn raid.

"How close were you guys to grabbing him before he hit the tunnel?"

"When we pulled up, they were still inside," Nico said. "I could see people upstairs in the window. Someone peeked through the blinds. By the time we got through that fuckin' door, he was gone."

I glanced at Brady, shaking my head in disbelief. "We knew he had a tunnel under a bathtub," I said. "We just didn't know he had one in every single safe house."

Lieutenant Zorro looked especially pissed at this remark. "No one has ever outrun me before," he said.

"We could hear them running—splashing in the distance—but had no idea where," Zorro said. "We found these lying in the sewer," he said, pointing back to two armored tactical vests, one black, the other a pale green.

Tucked in the vests were four black hand grenades with gold pins. One grenade had an American $20 bill wrapped around it. Chapo was presumably planning on tossing them behind himself to blow up the tunnel but hadn't had time. "He had the same setup in Cabo," I remembered.

Zorro handed me a red thumb drive that Chapo had dropped in the tunnel in his haste. There was not much on it other than surveillance video of the inside of some chick's house. Must have been another Guzmán obsession . . .

After a few minutes, we arrived outside Location One. There, an open brown garage door led into an enclosed driveway hidden by huge army-green canvas tarps hung from above.

In the garage, there was a small desk with several monitors displaying video footage from surveillance cameras in all of Chapo's safe houses. Someone obviously had the mind-numbing assign-ment of sitting in another cheap plastic chair in that empty garage, watching the small checkerboard of cameras on the screens.

This place was even older than Location Two—there were mid-sixties pink and green tiles throughout the bathroom, several bedrooms, and an old, filthy couch in the living room. The walls were bare. I had the feeling that this may have been Chapo's origi-nal safe house in Culiacán, considering its age.

Once again, a still-handcuffed Naris jimmied the hydraulic bathtub, revealing yet another entrance to a tunnel.

"Every safe house *is* connected," I said. They were close enough together that they could all be accessed through the same citywide sewer grid directly below the streets.

I walked back out to the street with Brady to get my bearings. It was now full daylight. I recognized the area.

"This is exactly where SEDENA killed El 50 back in August," I told Brady. "Right out on this street."

BRADY AND I HAD just jumped back in the old armored Suburban when I got an email from my group of intelligence analysts in Mexico City. Before we left La Paz, I had instructed them to ping other high-value cartel members every hour—anyone close to Chapo—so we could put them on the target deck in the event they needed to be located and arrested.

This latest email said that Picudo's pings appeared to have trav-eled at high speed, beginning in Culiacán and ending along High-way 15D just north of Mazatlán. I looked at the times.

Last ping before leaving Culiacán: 3:48 a.m.

Closest ping near Mazatlán: 6:00 a.m.

I reached over and nudged Brady.

"This is our drop!"

I knew that if Chapo trusted anyone, it would be Picudo. Guzmán might not want anyone else in the organization to know he was fleeing—in fact, the HSI team in El Paso reported that most of the DTO's lines were still up and running despite the chaos in Culiacán—but Picudo, his chief enforcer, could scoop him up and run him out of town discreetly.

"Yeah, looks promising," Brady nodded, studying Picudo's ping locations and times.

"This is our drop," I said again. "I'm telling you. Chapo's in Mazatlán."

"*Vámanos, güey!*" Brady said.

But we both knew that we couldn't go marching into Mazatlán with three hundred marines. Since the escape at Location Five, Brady and I already had Joe and Neil back in Texas scrambling for the next Top-Tier PIN to show up. Condor was bound to get a new device soon so he could get back in contact with Second-Tier and the office devices, making it appear that business was operating as normal. It was only a matter of time before El Paso cracked that new Top-Tier number.

"WHERE'S TORO?" I asked Nico. "We need to let him know that Picudo's on his way back now."

My intelligence analyst had told me that Picudo appeared to be on the highway toward Culiacán. I braced for the worst.

"He may roll in here with an army of guys," I said. "Ready for a fight."

"There's no place I'd rather be than right here," Brady said, looking around.

We now felt fully embedded with this SEMAR brigade, in the middle of the Sinaloa capital, and I could feel a subtle shift in the dynamic between us and the marines. Brady and I were no longer gringo federal agents with our mounds of intel and satellite imagery. It didn't matter how many narcos they'd hunted down; nothing had approached the intensity of this capture op. Zorro, for one, was impressed with the accurate intel I had been pitching for more than two weeks now. My precise pattern of life, coupled with all of the real-time intelligence generated by Brady and his El Paso team, had led us straight into Guzmán's lair.

Back at Location Five, I found Toro walking out of the kitchen, wearing a green-and-black *shemagh*. With his face wrapped in that camo scarf, he looked more like a US Spec Ops officer than a Mexican marine. To me, his Spanish nickname had a double meaning: he was bull-like, yes, but Toro also seemed to be short for *tormenta*— "storm." He'd been like a hurricane as soon as he touched down in Culiacán, tearing with ferocity through Chapo's secret underworld.

"Motor! Motor!" Toro yelled suddenly, calling for one of his young lieutenants so he could translate what I was about to tell him. Motor was only in his early twenties, but he was already a well-respected SEMAR officer, and he'd studied college-level English in the United States. He'd been part of the initial briefings Brady and I had with Admiral Furia and his brass back in Mexico City. I could normally get my point across to Toro in conversational Spanish, but for operational updates, Toro made sure Motor was there to translate, ensuring that no critical detail got lost or misunderstood.

"We need to hit Picudo," I told Toro, filling him in on the suspected drop. "Picudo will confirm that Gárgola's in Mazatlán."

"*Dale,*" Toro responded without hesitation. It meant "Let's hit him." "*Dale.*"

"Okay, by our pings, it looks like he's just coming into the city now," I said. "I'll get with Leroy and the marshals and put him next up on the target deck."

IN THE MEANTIME, Toro and his men were still squeezing Naris for more information.

"*Vamos a la casa de Condor!*" Toro said, moving toward the street.

Brady and I jumped in the backseat of Chapo's white armored Chevy Captiva; Toro had seized it back at Location Three and added it to SEMAR's fleet. My knees were jammed into the back of the driver's seat, but I was thankful now to be shielded by the armor on Chapo's former rig. Brady and I still didn't have any guns. Toro jumped in the front passenger seat and the Captiva sped off, following a dark gray Jeep Cherokee, another of Chapo's armored vehicles that the marines had commandeered. That vehicle had Naris inside—the Nose was leading us to the next takedown location.

I spun in my seat, looking back through the rear window at the long trail of *rápidas* racing through the streets behind us. I could hardly believe the pace at which the marines were smashing and grabbing, destroying Chapo's infrastructure.

We came to a quick stop on the rocky dirt road in front of a two-story concrete residence. The place looked like it was still unfinished. Stray dogs ran loose down the street while a young mother in skintight stonewashed blue jeans and black high heels walked outside with her young son.

"This is Condor's place?" Brady said. "What a shithole."

The marines were already inside, and while clearing the house they had found an old rifle and a photograph of a clean-shaven, light-skinned Mexican male with black hair tapering into a spiked flattop.

Brady studied the picture closely.

"Yeah, looks like a condor to me," he said.

Then we jumped back into Chapo's Captiva, once again winding through the city with the convoy of *rápidas*, and finally up a steep hill to a house in a much nicer residential neighborhood.

The moment I walked in the front door, I immediately noticed that the decor didn't match the bare-bones style of Chapo's other Culiacán homes. The furniture was far more expensive; the marble tile was shiny and clean; large framed artwork hung on the walls.

A mural just inside the front door was painted in deep shades of yellow, orange, and red. It was a memorial: I recognized the face of Edgar, Chapo and Griselda's slain son, ascending to heaven. I could almost hear Diego's voice, singing the lyrics of that narcocorrido years ago in Phoenix:

Mis hijos son mi alegría también mi tristeza
Edgar, te voy a extrañar

A white Chevrolet Suburban and Hummer H2 were parked in the garage, but it was suddenly clear that Naris had merely coughed up an old house belonging to Griselda, Chapo's second wife. There were no signs of recent activity—no fresh food in the kitchen, no dirty clothes in the bedrooms. In fact, it didn't look like she'd lived in the place for months.

"Regroup at Location Five," Toro called out.

The marines had dug up piles of photo albums in one of the bedroom closets, and before leaving Griselda's, I grabbed a stack of albums and tucked them under one arm.

When we arrived at Location Five, I walked upstairs and sat down on Chapo's brown faux leather couch. I peeled away the black balaclava from my face for the first time since putting it on, and only now did I begin to feel the first wave of exhaustion. I couldn't remember the last time I'd slept or eaten or drunk anything besides the few swigs of Johnnie Walker back at the base in Topolobampo.

Brady walked up the stairs with a couple of mugs of hot instant coffee he had found downstairs in the kitchen and handed one to me.

"Leroy's down there making eggs on the stove."

Brady and I began thumbing through Griselda's photo albums, trying to find any useful pictures of Chapo. But every family photograph of Griselda and her kids—Joaquín, Grisel, and Ovidio—was missing their father. Weddings, baptisms, quinceañeras, fiestas . . . but never a single shot with Chapo.

Once we were done with the photos, Brady and I scoped out the rest of the house. Next to a forty-inch TV on the living room wall, there was a second small white screen, the size of a large computer monitor, and downstairs by the small swimming pool we found the same setup: a forty-inch flat-screen TV on the wall and still another small white surveillance monitor mounted underneath it, showing pictures of all of Chapo's safe houses in the city.

"Anywhere he watches TV," I said, "he can keep tabs on

what's happening at all his houses." This was clearly one of the safe houses—La Piscina—where Chapo felt most at ease.

I walked back into Chapo's bedroom to take another look around and opened the closet, where I pulled down a black hat from the top shelf.

This was one of Chapo's famous plain, logo-free ball caps, which he could be seen wearing in the few verified photographs that existed of the kingpin since his escape from Puente Grande. Chapo always wore the black hat perched high on his head as if it were an essential part of his everyday uniform. I shoved the black hat underneath my bulletproof vest.

It was my only souvenir of the hunt.

SU CASA ES MI CASA

"ÁNDALE! APÚRATE! APÚRATE!" Captain Toro yelled.

We were still at La Piscina, but everyone was now scrambling to grab their gear and guns. Naris was giving up more locations. Brady and I jumped in the rear of the Captiva, Toro taking the front passenger seat again.

"Zorro de Toro," Toro kept yelling over the radio, giving directions in Spanish as the Captiva sped off, leading the convoy.

We raced into yet another residential neighborhood. SEMAR was battering its way through a door, and I walked in after the first wave of marines. The place was bare except for green bananas and pepinos (small cucumbers) that littered the floor, some white powder on the kitchen counter (cocaine cutting agents), and several black trash bags full of cultivated weed. I picked up a green banana from the bunch on the floor—it was a fake, used for international shipments of a far more lucrative crop, but they were all empty.

"How'd you like to be the poor fucker who had to fill each of these bananas with coke?" Brady said.

A single fake banana could hold no more than half a kilo of cocaine at a time—it would have been the most tedious, labor-intensive job. Immediately I remembered how Hondo, up in British

Columbia, was constantly looking for a warehouse large enough to store "fruit deliveries" for the boss. These fake bananas were most likely going directly to Vancouver to be unloaded, then shipped out to cities all across Canada.

A message from the El Paso wire room suddenly hit our group chat.

It was Lic-F reporting through Office-3, to Condor and Chapo.

Buenos dias, como amanecieron. En la ciudad siguen con alboroto esos del agua, no han dormido.

"Good morning, how did you guys wake? In the city, the rampage continues—those from the water, they haven't slept."

Lic-F continued:

Compadre andan bien bravos y todo el movimiento es contra la empresa.

"Compadre [Chapo], they are running really strong and the whole movement is against the business."

The convoy was indeed running on full steam again, smashing house after house.

We were now in another Picudo stash pad. In the dirt backyard, we found five fighting cocks, spurs attached to their legs, strutting around. They circled one another like prizefighters eager to spar. I watched as one dark red rooster with ocean-blue wing feathers attacked another. They were cocks trained to fight to the death. I picked up a handmade leather dice cup hanging on the wall, with markings burned into the side of it, a tribute to El 50.

By the looks of things, Picudo's crew had cleared out each stash pad just before we hit.

WE ROLLED BACK TO Location Three. The whole street was now blockaded by SEMAR *rápidas*. There was still the ever-present risk of a gunfight, but I was feeling a little more comfortable knowing we now had plenty of manpower out front.

Brady and I walked into Chapo's kitchen, looking for something to drink. I opened the fridge and grabbed the only three bottles of La Cerveza del Pacifico left.

"Split 'em?" I asked Brady and a few marines.

I smiled as I took one cold swig of Pacifico—remembering that night in Phoenix with Diego when I'd heard "El Niño de La Tuna," peeled back the canary-yellow label, and gulped down my first taste of the beer. I passed the bottle on to a young marine, then Brady took a gulp and passed it back. With the sleep deprivation, that little sip of cerveza was just enough. I let out a laugh—I felt like we were college kids passing around a bottle of whiskey.

I walked over to Toro with a newfound spring in my step.

"Vamos a tener otra oportunidad," I said—realizing I sounded a bit like my old high school football coach when it was halftime and we were down by two touchdowns. "We're gonna get another chance," I said, still holding the bottle of Pacifico. "This isn't over. I'm confident, Captain."

All we needed now was to obtain the new Top-Tier number. I explained to Toro that Chapo was smart enough to drop all of his phones, but Brady's team back in the United States was still intercepting several office devices, scrambling to intercept the new

Second-Tier so that we could identify the new Top-Tier device Condor was no doubt holding.

"My guys are on it," Brady said. "Just a matter of time."

"Until then," I told Toro, "we need to exhaust all our intelligence here, Señor."

"Bueno, vamos a Picudo entonces," Toro said.

The next target for us to capture became Picudo, Chapo's chief enforcer and the plaza boss of Culiacán.

SEMAR HAD BY NOW taken over every one of Chapo's five Culiacán safe houses, converting them into temporary bases. Brady and I climbed down the ladder underneath the bathtub in Safe House 3 to get a closer look at the tunnel.

It was the fourth one we'd seen that morning, and no different from the others except that it boasted a specially designed rack along the underground wall that had been used to store hundreds of kilograms of cocaine. The marines had found 280 kilos on the racks, along with boxes full of fake bananas.

I called Brady over.

"Check this out, man. Chapo's go bag."

Again, it was typical Chapo, purely functional: a plastic supermarket sack with a couple of pairs of white underwear inside. These were Chapo's favorites, those Calvin Klein briefs that Marky Mark had made famous. No toothbrush, no shoes, just those Calvin Kleins.

Brady laughed. "Shit, how many times do you escape down a fuckin' bathtub tunnel naked that you need to have a go bag full of tighty-whities?"

———

BY LATE AFTERNOON, still more SEMAR teams were out raiding Chapo's stash pads throughout the city.

One team drove back to Location Three in a white panel delivery truck with secret compartments built inside the walls and the bed. I watched as the marines pried open the traps and extracted fifty more kilos of meth. This time the drugs were packaged in plastic Tupperware-style containers with various-colored lids.

A little later, a crew of marine cooks arrived, bringing large steel pots and utensils, and took over Chapo's kitchen. I stood back and watched how SEMAR was settling into field life here in Culiacán: everyone would soon start receiving three meals a day, and the marines had even brought in a staff doctor. I still wasn't hungry, but I knew I needed to refuel.

Brady and I pulled a couple of plastic folding chairs up to Chapo's white plastic kitchen table and squeezed in next to Lieutenant Zorro. Zorro's warm smile and upbeat attitude masked his exhaustion. I watched, impressed, as Zorro skillfully cut into a tin can of scallops with his bowie knife.

"Cómo les gusta el campo?" Zorro asked. How did we like life in the field?

"Para mí," I said, *"me encanta."* I looked at Brady, still not quite believing we were sitting in Chapo's kitchen.

Spearing scallops straight from the can with the blade of his knife, Zorro reminded me of one of my uncles back in Kansas. He had the same genuine smile, the same rough-and-ready attitude of a born outdoorsman—he reminded me of someone you'd want to share a case of beer with, sitting around a fire, listening to his war stories.

In fact, I'd heard a few stories about Zorro already, from Leroy. El Roy had previously worked with Zorro on operations targeting the Zetas Cartel. One day Zorro had been caught in a fierce urban firefight; Zetas gunmen were raining hell on Zorro and his team from the rooftops, but Zorro walked calmly out into the open street, bullets whizzing around him, and methodically placed his troops into strategic shooting positions. El Roy said that in all his fugitive operations he'd never seen anyone so cool under fire.

My BlackBerry buzzed again: it was another message from the El Paso wire room. I read it and handed it across the table to Toro, who was sitting next to Zorro.

Condor, filtered down through Office-1 and out to Chapo's son Ovidio (Ratón):

La nana todo bien ai descansando oiga pero todo bien

"Everything is fine with Grandmother. She's here resting. Listen, everything is fine."

We knew that Nana was a code name Ovidio and Chapo's other sons often used for their father. It was a good sign that Condor was sending out reassurances that Chapo was settled someplace safe and that there was nothing for them to worry about.

LEROY AND HIS MARSHALS TEAM, Nico, Zorro, and a handful of marines had already left to track down Picudo. I sent them out near the Culiacán International Airport, knowing that, based on previous pings, Picudo most likely lived in a middle-class neighborhood near there.

The remainder of the marines and Brady and I all stayed back at Chapo's safe houses. We were resting for a few minutes while Leroy located the phone Picudo was holding. Once Leroy got a definitive address, he'd give the green light for us to mount up the convoy, rolling in fast and heavy; until then, we were able to briefly take a breather.

Brady and I were finally armed now, too—Nico had thankfully tracked down a couple of AR-15s from some marines and handed them over to us before he'd left.

"Damn, my net worth just quadrupled," Brady said, laughing, as he cradled the rifle. Since leaving Topolobampo, his only possession had been his BlackBerry. He'd even forgotten to grab his wallet before jumping on the helo with me, and joked that he didn't even have enough pesos to buy a toothbrush.

I WENT OUTSIDE to grab a breath of fresh air, spreading out on my back in the middle of Chapo's driveway and gazing up at the night sky.

The exhaustion hit me so hard, I felt like I couldn't move. It seemed as if the cold pavement was about to swallow me up. I called my wife at our La Condesa apartment, which wasn't the best idea, as I immediately started laughing and rambling.

"The night clouds, baby," I said. "The clouds of Culiacán. These are the clouds of Culiacán. The same clouds C would look up at if he was here and could look up."

"Where are you?" my wife said after a long pause.

"In C's driveway."

"What?"

"I'm just on my back, on the ground, looking up at the clouds. Remember how we'd look at the clouds when we were first dating?

There's one that looks like a gun! Where was that park we'd go to before the kids were born and just stare at clouds together for hours?"

I let out another burst of hysterical laughter.

"You're scaring me, Drew," my wife said. "You realize you're making absolutely no sense."

I had a serious case of the giggles. Brady came out to join me, and even he was starting to laugh.

"I'm serious, Drew—you really sound *messed up* . . ."

I realized, finally, that my nonsense rambling was alarming. I snapped to my senses.

"I'm *fine*, baby," I said. "I'm surrounded by some of the toughest warriors in the world. These marines are the best guys I've ever seen. I'm just—just kind of delirious. All I need is a good couple hours' sleep."

AND THEN I AWOKE with a sudden start. Somehow, I was in Chapo's bed.

"Luz verde!" a young marine yelled. *"Luz verde! Vamos a Picudo."*

Green light: we were going to snatch up Picudo.

It was still dark outside in the Culiacán streets. I rubbed my eyes. My head was aching, and I realized I must have nodded off for only forty-five minutes. As I sat up, Chapo's plastic garment bag peeled away from my sweaty back like a snake's dead layer of skin. The last thing I remembered was laying the plastic down beneath me on the bed so I wouldn't catch some STD.

I stood up, knees wobbly, stretched my arms high overhead and into a backbend. I'd been lost in a dream-memory of making that first big drug bust back in Kansas—three ounces of crack cocaine in that Deadhead's car. Was this whole thing a dream? Was I now actually sitting on a three-*ton* seizure of methamphetamine belonging to the world's biggest drug lord?

Brady and I scrambled to throw on our vests and slung the carbines across our chests. Once we were ready, we jumped into the Captiva with Toro and another young SEMAR lieutenant, nicknamed Chino—apparently for his Chinese-looking eyes.

We raced in another convoy out to a middle-class neighborhood near the Culiacán airport. Leroy had located the phone inside a small ranch-style house surrounded by a wrought-iron fence.

I looked at my watch: 1:32 a.m.

There wasn't another vehicle on the streets. The entire city was either scared or gearing up for war. I grabbed the pistol grip of my AR-15, pulling it tighter to my chest, as we turned the corner onto Picudo's dimly lit block. If there was going to be a Macho Prieto–style firefight, it was going to happen right here at Picudo's house.

EL 19

CROUCHING DOWN, I TOOK a position behind the front quarter panel of the Captiva, pointing my AR toward the dark shadows in the narrow walkway next to Picudo's house.

I could see in my peripheral vision a mob of black silhouettes quietly making their way to the front door. Then the stillness was shattered by the noise from the battering ram. Dogs began to bark.

The ramming continued for minutes. I was becoming more anxious by the second. Surely with Picudo's house under SEMAR assault, it would only be a matter of time before his reinforcements arrived on the scene.

The door was finally battered open, and I could see the stream of marines entering Picudo's house. I paused, expecting to hear a volley of gunshots.

Nothing.

Emerging from my position, I entered the house and saw three men on their knees in the living room, lined up against the wall.

I made my way through the kitchen and toward the back bedrooms.

SEMAR had Picudo at gunpoint on his bed. *No one*, I thought, *not even a feared cartel killer, looks scary when you roust them from a deep sleep, bare-chested, hair messed up, at 2:00 a.m.* Picudo didn't look like he could hurt anyone—he was pale, sweaty, and scrawny.

In Spanish, he claimed to be very sick. Brady didn't believe it; he grabbed Picudo by the left arm and spun him roughly onto his stomach. Picudo screamed now—a high-pitched wail. He claimed to be dying, but neither Brady nor I could decipher from *what,* precisely.

Brady pinned Picudo down on the bed. Now the SEMAR doctor came into the bedroom and told Brady to ease off; he wanted to examine Picudo.

"You gotta be shitting me," Brady grumbled to me. "They're gonna believe this pathetic actor crying like a little bitch?"

I listened closely as the doctor began asking Picudo the routine questions: How long had he been suffering from this rare disease of the blood?

Picudo exhaled with relief, righted himself in the bed.

Suddenly, Brady leapt forward—he had seen the butt of a gun secreted under Picudo's naked thigh.

Brady grabbed Picudo more roughly, pinning him on his face, holding him tight by the neck and the left arm.

"No lo toques!" the doctor shouted, "What are you doing? You can't touch him—he's very sick. He might die!" All the marines in the room were also yelling, shoving forward, agitated.

"Fuck that," Brady said. *"Tiene arma!* Look, the motherfucker's got a gun!"

Brady kept him pinned while one of the marines reached under Picudo's body and retrieved the Colt .45, fully loaded and with one in the chamber. If they'd believed this guy's bullshit, let their guard down long enough, Picudo could have pulled out his pistol and shot every one of us in the room.

I COULD STILL HEAR Picudo wailing as he went into the kitchen. There was a cache of automatic weapons on the table, including an AK-47, an AR-15, a TEC-9, and several other rifles. Picudo's men had been ready for a last stand—just like Macho Prieto's crew in Puerto Peñasco—but they had been caught by surprise.

By now, all the gunmen had been safely detained: cuffed, blindfolded, and lined up against the wall. The marines kept bringing more phones to the kitchen for me to analyze. The table was piled high with BlackBerrys and SIM cards, tossed in haphazardly with all the guns. There was even a book in Spanish that I was surprised to see: *La D.E.A. en México.*

Picudo—like Chapo—had been studying up on my agency and our operational history in Mexico. I had seen a copy already; this dog-eared paperback was well known in the DEA office back in Mexico City. It was a quickie knocked out by some writers for the *Proceso* magazine, using only a couple of retired DEA dinosaurs as sources, guys who'd been stationed in-country back in the 1990s. The biggest "revelation" was that DEA special agents operating within Mexico on counter-narcotics missions had been, illegally, strapped with guns.

I didn't care about the book or the weapons now, however: I hovered over the kitchen table, examining all the phones. I recognized some numbers that the Phoenix Field Division and Brady's people in El Paso had been intercepting.

Chino, the marine, led Picudo—now identified as Edgar Manuel López Osorio—out of the house, and he was strutting like he owned the city, which, as Chapo's plaza boss for Culiacán, he essentially did. There was no blindfold on him yet, and I was able to get a good look into those cold, steely eyes.

All I saw was an abyss.

Chino put Picudo in the back of a Jeep Cherokee, where he was joined by a fierce-looking marine—six foot four, powerfully built—whom everyone called "Chiqui" (slang for "the smallest"). Chiqui's face was pure Aztec—eyes dark and close-set—and his brow was pocked with scars. I'd never heard Chiqui speak, but it was clear to me that he was the muscle in this brigade.

"*Vamos,*" Chino said.

The convoy left in total darkness; I had no clue where we were heading. We ran down an unlit highway until, fifteen minutes outside Culiacán, we reached a pecan ranch—SEMAR owned it and had used it in the past for their training.

When we hopped out of the Captiva, I saw Picudo, blindfolded, grimacing as he sat on the Jeep's rear bumper, face lit up starkly by headlights.

There were now more than twenty people—various marines, Brady, Leroy, Nico, and me—surrounding the back of the Jeep. The darkness felt thicker now, the air electric: it was clear that Picudo was ready to talk.

His voice was a strong baritone, notable for its heavy Sinaloan accent. And the tone had changed from whiny back to that of a stone-cold killer. This was the *real* Picudo, the enforcer we suspected to be personally responsible for the murders of countless victims.

"*Mira, esto fue lo que pasó,*" he said calmly.

"Ándale," said Chino.

"Voy a estar honesto . . ."

The circle of marines undulated like some great jellyfish, growing tighter around Picudo.

"I'M GOING TO BE honest with you now," I remember Picudo saying. "When you hit the house, Chapo escaped through the tunnel— ran through the sewer. He was with a girl, Condor, and the cook. Chapo and the girl were naked—nearly naked. Just in their underwear. Chapo has a cut on his head from hitting something— running through the sewer. They called me to come pick them up. They escaped out of a drainage hole. When I entered the city, I saw all of your trucks."

Picudo had scooped up Chapo and his entourage in his truck and driven the boss at high speed down the Pacific coast. They drove for nearly two hours, and Chapo did not say a word the entire time, besides ordering Picudo to contact Bravo, Chapo's chief enforcer and plaza boss in the southern part of the state, and let Bravo know to meet them at the drop site.

"I dropped them off near the resorts," Picudo said finally. "I don't know where they went from there."

"What resorts?" Chino asked, glancing from Picudo over to me.

"Dónde? That's not good enough. *Where* on the coast?" Toro demanded.

Picudo's bared teeth flashed a cold hatred, his brow tensing behind the blindfold, before he finally gave it up.

"Mazatlán," he said, exhaling through pursed lips. "I dropped them at the *playa* exit."

"La salida de playa?" Chino repeated for confirmation.

"*Sí*," Picudo said. "Just before the new strip of resorts in Maz-atlán."

Turning our backs on Picudo, Toro, Brady, and I walked away to strategize.

"This confirms the drop," I said, "but we still need Top-Tier to know Chapo's exact location. Like I said, we need to exhaust all our intelligence while we're here in Culiacán."

Toro nodded his head in agreement.

"We still have all the sons and Lic-F to go after," Brady said. "They could give up where he's at, too."

"*Vamos a continuar,*" Toro said as we all climbed back into our rigs.

The convoy rolled out onto the dusty path. Then Chino pulled the Captiva over to the side. He was waiting for a couple of *rápidas* to join the end of the line. Toro turned in his seat to look back at me.

"*Que quieres hacer?*"

I felt like I was in a trance: I could see Toro's lips moving but couldn't make out what he was saying. The entire rig began to spin. I could feel the blood rushing out of my head. I was on the verge of passing out from pure exhaustion.

"*Qué sigue?*" Toro asked.

"*Dale!*" I replied, half-delirious. "*Dale! Dale!*" and I felt my fist smashing into my own palm.

"*Dale!*" Toro said, grinning.

Dale! Dale! Dale!

I kept repeating the word, so exhausted that my mouth was barely moving. Everyone in the car went quiet. I squinted over at Brady, whose head was tilted against the window. He was out cold. Chino was snoring in the front seat, and Toro's head was bobbing forward as he slept.

My eyes drifted shut.

A loud squelch came across the radio.

"Toro de Zorro! Toro de Zorro!"

We all snapped back awake, startled by the radio traffic. None of us realized we'd passed out and that the rest of the SEMAR brigade was waiting for us to lead the convoy back to the city.

I knew that we all needed some serious rest soon—we'd been running on fumes for days. The sun was just beginning to rise over the mountains in the east, and Toro made the command decision to head back to base—Location Three—so everyone could at least sleep for a few hours.

THAT AFTERNOON, Brady and I left Location Three and walked the dirt street down to Location Two, where SEMAR was processing all the evidence. We stopped along the way at a small *puesto*—a makeshift convenience store some guy was running out of his dark cinder-block garage. I bought a *paleta*—a Mexican popsicle—and a bag of Doritos for Brady with the few pesos I had in my pocket.

When we walked into the Two, I saw that SEMAR had extracted all the meth from the tunnel. Brown packages were stacked on top of one another and covered the entire living room floor. In the kitchen, a young marine was counting plastic bananas and placing them in a large container.

I stepped outside, passing the stash of rocket-propelled grenade launchers, AK-47s, and other military-grade weapons, laid out in meticulous order on the white pavement.

Something shiny caught my eye—it was the gun Chapo had ordered Naris to fetch for him. The Colt Super .38-caliber Automatic had Chapo's initials monogrammed into the grip, in diamonds: J.G.L.

Despite my still-exhausted state, I now knew I wasn't in some waking dream: holding that cold steel in my hand made everything tangible.

The detailing on the Colt Super was impressive. Chapo hadn't had time to grab the pistol before he fled through the tunnel and sewer, and it was clear that this was his favorite weapon, his personal Excalibur.

Who knew the full history behind Guzmán's pistol? But if the Colt gave Chapo some mystical power, I could almost sense it now, too. Just holding the .38 Super in my hand, I felt that same visceral energy transferring through my grip.

Stacks of BlackBerrys were piled high in the backyard. Brady and I sat down and began looking through Picudo's phones one by one. I found a picture of Duck Dynasty and messages in the most recent chat logs with Lic-F.

Just then, another message hit our group chat from El Paso. It was a fresh intercept—Lic-F reaching Condor and Chapo:

A poco tuvo problemas el picudo

"Oh really, Picudo had problems?"
Condor and Chapo immediately responded:

Si. Tenemos ke estar trankilo. Por ke. No keda de otra. Claro. Por ke picudo. Pobre. El si sabe de todo.

"Yes, we have to be calm because there's no other option. Yes, of course. Because of Picudo. Poor Picudo. He knows everything."
I called in Nico and Leroy.

"We need to find Lic-F next," I said. "Now that Picudo's gone, Chapo's going to be relying on him for everything."

"Agreed," Leroy said. "We'll focus on him and the sons."

Then my BlackBerry buzzed with another message from El Paso. It was Condor to Chapo's son Ratón:

oiga dise inge si tiene una super. Ke le mande. Con 4 cargadores estra. Es para el oiga. Y si me ase el paro ai oiga con 1 bereta o lo ke tenga oiga

"Listen, Inge is asking if you have a Super [Colt .38] that you can send. With four magazines. It's for him, and do me a favor: bring me the Beretta or whatever you have."

Then, moments later:

oiga dise inge para kele mande 10 rollos al negro.

"Listen, Inge says to send ten rolls to Negro."

I grinned. I knew that "Negro" was another code name for Manuel Alejandro Aponte Gómez, a.k.a. "El Bravo." Chapo needed ten rolls—$100,000 in cash—delivered to him immediately. This was confirmation of Chapo's vulnerable position: he was free, but he had nothing with him in Mazatlán—no guns and no cash.

We went back to analyzing the BlackBerrys, and, as ever, minutes turned to hours.

I vaguely remembered the marines handing out sandwiches for dinner, but when I glanced at my watch, I saw that it was now one o'clock in the morning. I lay down on the top of Chapo's bed, the mattress now covered only by the dirty brown fitted sheet. Brady was sitting in the corner on a chair. I stared at the ceiling, imagining where Chapo might be resting in Mazatlán.

"Weird, isn't it?" I said. "Chaps is somewhere right now in Maz trying to figure out his next move, and we're here in his bedroom strategizing ours."

"Good to know he's got no money with him," Brady said.

"We need that Top-Tier."

"We'll get it."

"Yeah," I said. "Soon enough."

We were both confident that Joe and Neil, in El Paso, were dialed in and cranking out the roving intercepts as fast as they possibly could.

"*Luz verde! Luz verde!*" one of the marines yelled down the hall. El Roy had locked down on the phone of Iván Archivaldo Guzmán Salazar, in a house on the north side of the city.

So much for rest—Brady and I jumped in a *rápida* with Admiral Garra and raced to the location. By the time we arrived, SEMAR had already made entry, but Iván was nowhere to be found. Instead, there were only a couple of his workers sitting on a stash pad full of weapons, two-way radios, a small amount of meth, and, of course, another pile of BlackBerrys. Iván had clearly been warned by Chapo to get out of the city and had set up a classic BlackBerry mirror operation with his workers before leaving.

Nico handed me a black baseball cap he had found in one of the bedrooms. The front of the hat was emblazoned with #701 in shimmery gold stitching—again, that *Forbes* ranking of the world's wealthiest men.

As we continued to look around, I received a new email from my analysts back in Mexico City. A brand-new Nissan GT-R

belonging to one of Chapo's sons, Jesús Alfredo Guzmán Salazar, had just pulled into the Mercedes dealership off Boulevard
Pedro Infante.

"*Vamos!*" Toro said. "Do you know where the dealership is?"

"Yes, of course," I said.

"All right, you lead, then."

Brady and I jumped into our old armored Suburban and took
off out of the neighborhood, trailed by a stream of *rápidas*.

WE TOOK THE MERCEDES dealership by storm. Guns drawn, the marines flooded the showroom floor and the service center and surrounded the parking lot. Brady and I rushed inside, looking for
Alfredo, a skinny twenty-six-year-old narco with a baby face.

Alfredo's Nissan had arrived at the service department no more
than five minutes before the convoy. I put my palm on the hood:
the engine was still hot. The GT-R had a temporary California
registration sticker on the windshield—further proof, I knew,
that this was all part of Alfredo and Iván's long-running money-
laundering scheme. Chapo's sons would send a worker in the States
to pick up hundreds of thousands of dollars in drug proceeds and
then "smurf" the cash into various US bank accounts—making
multiple deposits of just under the $10,000, the federal manda-
tory reporting requirement. Once the cash was in the US banking
system, Iván and Alfredo could use aliases or straw purchasers to
negotiate the best price for these exotic sports cars. Their workers
in the United States would wire the money to the seller and arrange
to have the car imported into Mexico and delivered straight to
Culiacán.

Brady and I bounded upstairs and cased the executive offices, but there was still no sign of Alfredo. By now the entire dealership, inside and out, was crawling with armed men in green-and-black camouflage. All the employees and customers were in shock—no one dared speak a word.

We reviewed the surveillance video of the past hour. Then I found Captain Toro. "Alfredo's not here," I said.

"He's not even on the surveillance tape," Brady said. "He had a whole crew of narco juniors dropping these cars off."

Brady pointed to the string of brand-new Mercedes sedans and coupes lined up in the service area. Captain Toro took a few moments to review the video and returned to the shop floor.

"We're taking them all," Toro said, and he began walking through the parking lot with a group of marines, checking every vehicle. "If it's armored, we're taking it."

Fourteen armored vehicles were seized, and six more luxury cars, even a Ducati motorcycle. As evidence, I began snapping photos of the makes, models, and license plates. Mercedes SLS AMG. Mercedes AMG G63. Mercedes C63. Mercedes CLA45. Even a cloned municipal police armored Dodge Charger.

Chapo's son Iván had the most expensive car on the lot: his 2010 silver Mercedes-Benz SLR McLaren two-door coupe, complete with suicide doors, a customized sound system, and a 5.4-liter supercharged V8 engine. Chino popped one of the batwing doors and fired it up. The McLaren sounded louder than a Learjet.

Brady and I got back in the Suburban and watched the marines hop into all the cars, driving millions of dollars of vehicles right off the lot, one after the other, as though they were just playing a wild round of *Grand Theft Auto*.

AS WE STARTED BACK toward Location Three, in my rearview mirror I could see the Mercedes dealership employees standing outside gawking, still in shock.

At that moment I realized that we had taken full control of the city—we'd wrested Culiacán away from Chapo. The SEMAR machine was untouchable; no one in El 19 had had the balls to confront us. The marines were moving too fast and hitting too hard. All of Chapo's *halcones* had crawled back into their holes. Even Lic-F was reporting intel from his corrupt sources that was stale—two hours old.

I thought back over the past few days. I couldn't remember seeing a single police car—local or federal—patrolling the streets. All law enforcement officers were now obviously in hiding. Even the city's dirtiest cops knew that it was best to stay out of SEMAR's path.

BACK AT THE THREE, the entire block looked like a bizarre luxury car show in the middle of the hood. The immaculate Mercedes sedans and coupes were parked bumper-to-bumper on the dirt-and-gravel street.

Another message came fresh from the El Paso wire room.

Lic-F to Chapo and Condor:

Por otra parte hay nos sacaron unos carros duros de la agencia esos del agua, y andan duros aun.

"On another note, the ones from the water took some hard [armored] cars belonging to the agency [Chapo's DTO]. They're running hard."

Condor answered almost instantly.

Buenas tardes sr. Dise su compadre kesi los carros eran suyos. O los menores.[*]

"Good afternoon, sir. Your compadre is asking if the cars were yours or the minors."

Lic-F replied:

Unos duros eran mios, pero sacaron otros de lujo que yo creo si eran de los menores.

"Some of the [armored cars] were mine, but they took other luxury ones that I believe were the minors'."

This was confirmation that Chapo's DTO was using the Mercedes dealership as a ruse, a place to store their most prized possessions so they wouldn't be seized by SEMAR as they tore through the city. It was also clear that they thought they could still weather the SEMAR storm.

Leroy and Admiral Garra were standing in the driveway at Location Three.

"We need to go after Kava," I told Garra. "He can tell us where every single lightbulb is, where every trapdoor and secret passage is tucked away. He constructed all the tunnels on every piece of property Chapo owns. If we're going to destroy this place, giving him nowhere to return, then we've got to find Kava—he'll give us everything."

---

[*] *Los menores*—literally "the minors" or "youngsters"—was frequently used within the DTO to refer to Iván and Alfredo.

Leroy, Nico, and the marines hit the streets again, carving circles within the city, trying to locate Kava's phone. But they had no luck, and by now everyone was at the point of exhaustion.

The clock was ticking, and Admiral Garra was getting stressed again. The Mexican Attorney General's Office was taking over all the locations in Culiacán, booting us all out of the safe houses and stash pads we'd been using as our makeshift bases. Rumor had it that they wanted to begin filling up all the tunnels beneath Chapo's safe houses with concrete.

"You'll never keep this guy from going underground," I told Admiral Garra. "He's like a mole—he'll try tunneling again in no time—trust me."

Garra said they might have no choice but to wrap up the mission soon.

I shook my head.

"Tenemos que mantener la presión," I said. We couldn't let up with the pressure. I reassured Garra that I was still confident we'd get Top-Tier soon.

"We're almost there," said Brady. "One more day—tops—we'll have it."

Chapo simply couldn't function without his communications in place.

"We just need a bit more time, sir," I said.

"More time?" Garra said somberly. "That's the one thing I can't promise you."

MIRAMAR

"PACK YOUR BAGS, GUYS—ROLLING OUT!"

It was Chino, shouting from the doorway. We had to vacate the safe house. It wasn't hard to snatch up our possessions—I could carry everything I had in one hand: a leather laptop case containing one MacBook, and a few phones. Brady had just his Black-Berry. Neither of us had changed clothes or underwear in more than a week.

"I can feel my shirt rotting off my skin," I said.

But I didn't like that we were leaving home base; it was just beginning to feel secure. Chapo's place had become our safe haven.

Su casa es mi casa.

I realized I was going to miss the camaraderie, not to mention the marines' home-cooked meals. One thing I wasn't going to miss: the bathroom with the sign Chino had duct-taped on the door: EXCLUSIVO CAPITANES Y OFICIALES. Every morning, marines lined up by the dozen, waiting to use a filthy toilet that lacked both a seat and toilet paper.

Brady and I jumped into an armored Volkswagen Passat—another of Chapo's customized cars. No orders had been given on a destination, but I could see that it was deeper into urban

Culiacán. Eventually we passed a water park and drove onto the city's main baseball field. The well-groomed lawn quickly filled with all the glossy Mercedes, mixed in with the mud-caked SEMAR *rápidas*.

"A baseball diamond?" Brady said, laughing. "Sleeping in the open air?"

"May not be that crazy," I said. "It's probably the safest place to be in the entire city. At least we can control the perimeter and see everyone coming and going."

Out of nowhere, a dented, rust-flecked white pickup pulled up outside the fence. *Chapo's halcones?* I nudged Brady, glancing in the direction of the truck.

"Fuck," I said. "Here we go."

Why would any outsider be approaching a field full of heavily armed marines? I instinctively looked for a place to dive; there wasn't much cover besides a few shabby-looking trees.

"Dude, look at all those cots," Brady said, laughing.

As the truck got closer, we had realized that the flatbed of that old truck had been stacked fifteen feet high with rudimentary military beds, constructed of wood and strung with potato sacks. It sure would be a step up from sleeping on the cold tile floor without blankets in Chapo's safe houses.

As the sun set on Culiacán, Brady and I went to find Admiral Garra and Captain Toro. We didn't want to be overheard, so we met in the growing darkness behind the concession stand at the baseball field. Leroy, Nico, Chino, and another young SEMAR lieutenant, Tigre, were there, too.

"What's the latest intel?" Admiral Garra asked.

"Gárgola's instructed Lic-F to find two houses for him on the coast. We're still waiting for Top-Tier, but I think we need to move down to Mazatlán. Set up shop at a resort and begin working our intel there."

"We need to get down there before he has a chance to flee," said Brady.

Admiral Garra nodded, then gave us the bad news: Captain Toro had to leave Culiacán immediately—his brother had been struck in a hit-and-run accident in Mexico City and it didn't sound like he would make it through the night. With Toro leaving, front-line command would be in the hands of Chino and Tigre.

"We've got about two more days before we need to wrap this up," Admiral Garra said. "Then I'm pulling everyone out of Sinaloa."

We all agreed it was best to move to the Mazatlán resort strip and continue working from there.

But it was crucial to avoid all tails and countersurveillance.

"We can't all go down there in a convoy," Chino said.

"You're right—none of these *rápidas*," I added. "Gárgola's people will spot them the second we leave town."

"Agreed," Brady said. "We leave all the SEMAR vehicles here. Make a covert approach to the south, taking different routes."

"We'll use all of his own *blindados*," Chino said.

What better vehicles to use than the fleet of Chapo's own armored cars and trucks we'd seized?

Under cover of darkness we drove over to Soriana—a Target-style chain popular throughout Mexico—still wearing our camouflage fatigues, boots, and black balaclavas.

Brady, Nico, Leroy, a few marines, and I spent an hour loading up on sleeping bags, toothpaste, shorts, shirts—and the first fresh socks and underwear I'd seen in weeks. Brady and I would have to look like typical Americans on vacation, so we also grabbed the most basic red-and-black T-shirts, baggy boardshorts, and flip-flops.

The Soriana shoppers stared at us like we were nuts. I realized how out of place we looked, as if we'd just parachuted in from Iraq . . . Or maybe we resembled a couple of narcos come to kidnap someone in the store. One of the customers, a middle-aged woman, stared into my eyes. Then she cracked a smile: she must have realized they weren't the eyes of a narco hidden behind my black mask . . .

Back at the baseball diamond, we grabbed platefuls of tacos al pastor just outside the fence, and the marines let in a kid on a banana-seat bike with a cooler full of fresh tamales con pollo, fifty pesos each.

With our bellies full, Brady and I walked toward a large open-air room with bare orange-painted steel pillars and screens to keep the bugs at bay, filled with tightly spaced cots. The warm wind was wafting through as I sprawled out on my cot.

"God, these potato sacks are better than a Sealy Posturepedic," I said.

I unlaced and kicked off my boots. It was the first time my feet had been out of those sweaty things in five days, and my big toes had water-filled blisters.

In a brand-new black T-shirt and BDU pants, barefoot, I took one last stretch and muttered two words under my breath: "The blind . . ." Delirious, I shut my eyes and was instantly more than a thousand miles away.

I LOVED LIVING right on the river—there was no fence separating our backyard from the water's edge—and on warm days, my brother,

Brandt, and I would wade out to a nearby island, claiming it as our personal playground, building forts out of sticks and poking around in muskrat dens.

It was late fall—I was ten years old—when our father told us that we could both come along on a hunt. We'd been preparing for this day since we were toddlers, walking around the living room blowing old wooden duck calls our father handed down until our mother would beg us to stop. Whenever our dad came home from a hunt, we'd help him unload the pile of ducks from the flat-bottomed, sixteen-foot PolarKraft and toss the plastic mallard dummy in the yard for our black Lab, Rough, to retrieve.

The night before the big day, Brandt and I had both been so excited that we'd crawled into our bunk beds all ready, wearing our brown camouflage jackets and face paint. At 5 a.m., our father snapped on the bedroom light, and we sprang out of bed and threw on our black snow pants and gloves. It was still pitch-black outside, and I clutched my brand-new Remington shotgun mid-barrel as we walked together over the frost-covered lawn and down to the forest-green boat tethered at the river's edge.

The ride upriver was ice cold and wet. My ears were numb, but I didn't reach for the stocking cap my dad had given me. He didn't wear one—why should I? The cold wind whipped across my face. The heavy steel boat pushed through the water, the front edge of the hull splashing cold waves up over the sides, soaking my jacket and catching me on the side of the face. I caught sight of a group of ducks splashing out of the river, kicking up near the bank.

My father cut the twenty-five-horsepower Mercury motor and the PolarKraft drifted on in silence. Under the moonlight, I felt as if we'd been floating in the middle of that river for hours, my father flashing the fifteen-million-candlepower spotlight along the shore, trying to locate the well-hidden duck blind.

I saw the outline of the blind that Brandt and I had helped build out of wood, camouflaged with cattails, downed tree limbs, and other foliage. We threw out the decoys one by one so they floated near the front of the blind.

With my Remington loaded, I sat on the five-gallon bucket, peering through the narrow open slots between cattails.

The sun began to streak the horizon in shades of orange and pale gold through the trees across the riverbank. I was startled by the sound of whistling above my head: faint at first, trailing off in the distance. I looked up but couldn't see anything. My father pointed up to the sky.

"They'll be back."

Eventually we heard the whispering whistle above us once again, and this time we spotted the flock, their wingtips catching the faint rays of the sun. But once again the ducks disappeared.

I picked up my duck call and gave a couple of quacks through the double-reed lure. Brandt and I took turns letting out the sounds we'd been practicing since we were old enough to walk. The noise of the calls soon filled the river valley.

"Get down—they're coming back," my father whispered. In a few moments, the ducks were out in front, circling over the tops of the floating decoys. "Let them come in close," my father said.

My leg began to shake as I tried to keep myself still. I could see the flock out in front starting to descend rapidly, wings locked, a couple of the mallards dropping their shiny orange legs like the landing gear on a jet.

"You ready, Drew?"

I was silent, just a quick nod to my father, the ducks no more than thirty yards away now. I could make out the colorful green heads and bright yellow bills of the drakes.

"Take 'em!" my father yelled.

I quickly stood up and shoved the Remington deep into my shoulder. I saw nothing except that green head hanging in front of my barrel, the wings seeming to flap in slow motion. My right shoulder jerked back and the yellow plastic shell flickered in the corner of my eye as I ejected the first round. *Miss.* I swung the gun slowly, tracking the bird as it flared across the sky, and pulled the trigger again. *Miss.* The duck kept flying.

Last shot.

The Remington held only three shells.

Start behind the bird, pull through the bird, and fire. I repeated the precise words my father had taught me like a catechism. The mallard was gaining distance quickly, just beginning to exit the kill zone, when I gave one last, slow squeeze of the trigger.

"Dead bird!" my father yelled to Rough, who was panting eagerly at the blind's edge. "I thought he was gone, Drew, but you stayed on him . . ."

AND WITH THAT LAST SHOT, I awoke with a start in Mexico, still thinking for a moment that I was in that Kansas duck blind, feeling a wet scraping on my bare cheek.

"What the hell?"

It was bright daylight, and a little dog was licking my sweat.

I wiped the sleep from the corners of my eyes, wondering where this puppy had come from. It was a blue-eyed husky with a red collar and a spherical bell around his neck. He was scampering around in the bright sunshine, sniffing and licking everyone.

Chino told me that a couple of the younger marines had found the dog alone in one of the safe houses—with no food or water—and had brought him along as the new team mascot. Someone had given it a new name, too: El Toro, in honor of our missing street commander.

Brady and I quickly packed our bags full of all our Soriana gear and jumped in the backseat of Chapo's Volkswagen Passat, now wearing our new T-shirts and boardshorts. A young marine lieutenant took the wheel while Chino hopped in the front passenger seat.

"This is the perfect ride," I told Brady—it was low-key, without any of the bells and whistles of the typical narco car.

Chino stopped at Plaza Fiesta to grab a few last-minute supplies; I immediately recognized this as the spot where Chapo would always send his people to be picked up by Naris when he wanted to confer with them face-to-face.

Brady and I walked into a small mercado and bought a plateful of rolled taquitos topped with queso fresco and salsa verde while we waited for Chino to finish up.

Seeing me weaponless, Tigre had lent me his FN Herstal Five-Seven pistol, a Belgian-made semiautomatic. It was a small-caliber gun—firing rounds of 5.7 by 28 millimeters—but effective at close quarters: the rounds could penetrate a bulletproof vest, giving the FN Five-Seven its street name: "the cop killer."

Brady and I were shoulder-to-shoulder in the backseat when both of our BlackBerrys buzzed with news from Texas.

Joe and Neil, in El Paso, working with Camila and her team of assistant US attorneys, had done it: the roving wire they'd taken so long to write and get authorized had finally hit pay dirt.

"New Top-Tier!" Brady shouted.

"Yeah, baby! Condor's up and running," I said. "And the prefix six-six-nine."

"Yep, six-six-nine."

All of Culiacán numbers had a prefix of 667. This prefix, 669, I immediately knew meant the phone was from Mazatlán. I flipped open the screen of my MacBook, balancing it on my knees in the back of the Passat, and hit the ping button. Within seconds I had a hit. The device was active, right along the beachfront strip of resorts.

Someplace called Miramar.

Hotel Miramar.

THE MAN IN THE BLACK HAT

LEROY HAD LEFT CULIACÁN with Zorro and his crew an hour earlier and was already arriving in Mazatlán. I sent him the new Top-Tier number.

"El Roy's headed there now, near the hotel, to confirm the ping," I said to Brady.

"I just hope Condor keeps it on long enough," Brady said.

This was it—we were on our way down to the water for the last shot.

In that cramped backseat, I felt my leg beginning to shake. I was becoming more impatient by the minute. *"Ándale!"* I yelled up to the front, slapping the young lieutenant on the shoulder. The engine revved as we accelerated, but the heavy Passat still felt like it was crawling its way to the coast.

My BlackBerry buzzed with a new message from Leroy.

"Confirmed. I've got it at Miramar."

SEMAR had rented a small house on Calle Bernardo Vázquez— a private home in a sleepy residential section of Mazatlán—so we could set up our base of operations discreetly, away from any of Chapo's *halcones*.

When Brady and I reached Mazatlán and walked into the

house, we nearly tripped over all the piles of tactical gear littering the floor. Everyone was in high spirits. Loud laughter boomed in the living room, and someone had just ordered a pizza. Several marines were lounging around on the couches, watching TV, and a few more were sitting around the kitchen table with Leroy and his team of marshals.

Leroy got up and motioned for Brady and me to follow him to a quiet corner of the house.

"How confident are you that Chapo's with this Top-Tier device?" Leroy said.

"One hundred percent," I said.

"How can you be so sure?"

"Condor types most of the messages," I said. "But sometimes Chapo picks up the BlackBerry and types them himself."

"How do you know?"

"Chapo spells like a kindergartener, he doesn't know how," Brady said.

"Like this message—just came in an hour ago." I handed my BlackBerry over to Leroy and showed him the screen. "Chapo's talking about a house he's planning to move to. Look at the spelling."

Sy pero no tyene pura kosyna mannan en la mana le pone mynysply

I translated aloud.

"'Yes, but it doesn't have a full kitchen. Tomorrow morning he'll put in a mini . . . supply'? Mini-something—who the hell knows what he's typing there? But you see the way he spells 'kitchen,' *cocina*?"

"Yeah, '*kosyna*,'" Leroy said, nodding.

With Chapo's writing, I explained, there was a consistent substitution of *y* for *i* and *k* for *c*. Guzmán would write *bien* as *byen* and *cuanto* as *kuanto*. This wasn't typical Spanish texting slang or shorthand; it was unique to Chapo. He typed virtually every word phonetically. And his messages were peppered with self-taught constructions. Even as elementary a word as *caliente*—Chapo spelled it *kalyente*. These were clear forensic tells—Chapo, not his secretary, was fat-fingering these messages.

"So you're positive he's in the room with the device we're locating," Leroy said.

I grabbed the one war trophy I'd been carrying and pulled the black hat down tight on my head.

"Yep, I'm positive, El Roy. An hour ago, when this message was written—*kitchen* spelled like that—the BlackBerry was directly in Chapo's hands."

The pizza had arrived, and everyone was grabbing slices, but I had no time to eat. My bosses back in Mexico City had arranged for us to use three of DEA's armored Suburbans—this time brand-new ones, the best in our fleet. I left Nico in charge at the house.

"When Tigre arrives," I said, "get with him and come up with a plan for the takedown. Brady and I need to run into town and grab these rigs. When I get back, we'll finalize everything for the capture."

"You got it," Nico said.

IT WAS 12:30 A.M. when Brady and I walked back into the house. Every single one of the young marines was fast asleep. Leroy and his guys were passed out on the couch and floor, too—there were no blankets or pillows, just a bunch of bodies sprawled out on the bare tile. Even Nico had crashed upstairs in one of the beds.

"I know everyone's tired," I said to Brady, "but c'mon, for fuck's sake . . ."

Here we were, about to go grab the world's most wanted man, and everyone was racked out?

"Get up!" I said, shaking Nico. "How'd the meet go with Tigre?"

Nico opened his eyes, still half-dazed.

"The meet with Tigre," I repeated. "What was his takedown plan?"

"He never showed up."

"What do you mean?"

"He never showed."

"Where the hell is he?"

"He's staying with a group of his guys at some shitty motel on the outskirts of town."

"Fuck!" I shouted. "If he has a plan, what good is it if *we* don't know it? Get up, dude—we need to find him now."

Brady and Nico jumped into the Suburban, and after a twenty-minute drive I snaked the vehicle back through several winding alleys, screeching to a stop near the motel's check-in office.

"Look at this dump," said Brady. "Red neon lights and garages."

This was the kind of place the locals took their hookers to for an hour or two. Each room even came equipped with a garage so you could discreetly hide your car for the duration of your stay.

"What room is Tigre in?" I asked Nico.

"Don't know," Nico said. "He's not answering his phone."

"Let's just start banging on the fucking doors."

I was already beginning to miss Captain Toro: now we were rolling with a cadre of very young marines—all in their twenties—

full of energy and experience but lacking Toro's coolheaded focus and leadership.

We needed to communicate, coordinate, lay out a thorough takedown plan—one that allowed for any contingency or screwup. I felt like everyone was running a little on the wild side. This was too improvised.

We all split up and started knocking on doors. We startled two shady-looking locals and woke up groups of groggy marines, sharing tiny rooms, trying to grab a couple of minutes of shut-eye.

We found Tigre in the last room of the motel. We'd clearly woken him from a deep sleep, but he was awake enough to take us into the adjacent garage, where we could speak more privately.

"*Carnal,*" I said. "If you have a plan, we don't know it."

"Of course we have a plan," Tigre said, shaking off the cobwebs. "We've done this many times before."

"Tigre, I'm worried about our manpower and perimeter," I said. "And why is Chapo holed up at the Miramar? I'm sure he knows every floor plan, staircase, and exits to the street; we don't know any of it. How many guys do you have?"

"I've got forty marines here," Tigre said. "We'll flood the hotel and put a couple of *rápidas* on the perimeter—"

"No, that's not enough!" I cut him short. I realized that to Tigre this was just another hit, another door to smash. He and the other marines were almost numb to it now; they'd been doing these hits day after day in Culiacán, and predawn raids had become routine.

"We need more men on the perimeter," I said. Brady nodded in agreement. "And as many guys as you can get inside."

"We've got another brigade down the street," Tigre said. "As soon as we get the green light, I'll call them in."

"How many men?"

"I'll have another thirty marines in fifteen minutes. Then thirty more after that."

"Fine," I said. "Sixty extra men should do it. And where are the helos? We're gonna need air support in case he manages to escape the perimeter."

"The helos are two hours away," Tigre said.

"No, that's not gonna work," I said. "We need them closer."

"I'll move them down to Culiacán. Once we give the green light, it will take them an hour to fly down."

"Perfect," I said. "Stage them there. We don't want them any closer than that. The movements may spook him."

"*Claro*," Tigre said.

"Show me where you're placing *rápidas* on the perimeter," I said, pointing to the Google Earth view of the Hotel Miramar on Tigre's iPad. Tigre said he had only three *rápidas* for the capture op.

"Only three?" I asked. "How do you see us moving in?"

"We'll use your vehicles and fill them with my guys. We'll ride right up to the front gate of the hotel and enter from there."

"Fine," I said, finally exhaling a bit.

"Meet back here at oh-five-hundred hours, ready to go," Tigre said.

IT WAS JUST AFTER 3 A.M. when we left Tigre and headed back to the rental house. Nico and Brady walked upstairs to rest.

I was too wound up to sleep, and everyone had to be awake in an hour anyway. I sat down at the kitchen table and studied the block around the Hotel Miramar on my MacBook over and over. I didn't want to leave a single detail of the op to chance. We'd have the front entrance covered, but I was still worried about Chapo slipping out a back or side door and into a vehicle on Avenida Cruz Lizarraga, behind the hotel.

What if he'd worked something out long in advance with Kava? What if they'd constructed one of their hydraulic tunnel entrances in a room on the ground floor of the hotel? Or had some other way to access the sewer system directly from the hotel basement? A manhole or drain somewhere out in the street?

I'd been wearing Chapo's black hat so long that the brim was sweaty and sticky, and I could feel my forehead starting to break out. Finally I grabbed a slice of pizza and quickly typed an update for my group supervisor back at the embassy in Mexico City.

2/22/14, 3:33:05 AM: ANDREW HOGAN TO GROUP SUPERVISOR
███████████████: 23.226827-106.427935 targ loc hitting door at 0530—he's there

I hit SEND, took a bite of pizza, and felt my anxiety slowly dissipate. I even managed to crack a smile at my reflection in the screen of the laptop, my sleepless eyes beneath Chapo's crazy-ass black hat.

My thoughts turned to Diego, probably fast asleep back in Phoenix. I knew my old partner would have given anything to be right there in Mazatlán, prepping to make the predawn capture raid on El Niño de la Tuna . . .

Cuando nació preguntó la partera
Le dijo como le van a poner?
Por apellido él será Guzmán Loera
Y se llamará Joaquín . . .

I started to text him but stopped midsentence: no need to wake him.

I looked down at the clock on my laptop. It was 4:00 a.m. sharp.

"Despiértate!" I shouted, getting up from the kitchen table. *"Despiértate!"* I walked around the house, yelling for everyone to wake up, flipping on the lights and ripping off blankets.

"It's go time, guys! Up! Up! Get up!"

PITCH-DARK: 4:58 A.M. I took a deep breath and jumped into the driver's seat of the white Chevy Suburban, now loaded up with marines who were all armed with AR-15s—Tigre rode shotgun. I glanced in the rearview mirror at Brady, who was in the driver's seat of another white Suburban, loaded up with his own crew of marines—together we formed the entry team, with Tigre and me leading the way down to the strip of hotels and condos on the Malecón, the thirteen-mile-long boardwalk of Mazatlán.

We sat, waiting for the green light from Leroy, Nico, and their SEMAR crews. The phone-finding and security teams were looking for final confirmation that the Top-Tier BlackBerry was indeed still at the Hotel Miramar.

Sitting in the idling Suburban, I typed out a quick message to my father back in Kansas.

"Going in hot."

Just then, Tigre's radio squelched. I heard the words we'd all been waiting for:

Luz Verde.

"Vamos," Tigre said. I threw the Suburban into drive, ripped out of the parking lot, and took off down the desolate highway.

No one in the rig said a word. All the marines were quiet, checking their guns, focused on the mission. En route, we were joined by three *rápidas*, and together we sped in a tight convoy down Mexican Federal Highway 15 into the heart of Mazatlán.

In just under eight minutes the convoy was on the Malecón, but as I went to make a left turn onto Avenida del Mar, I was blocked by a squad car from the municipal police department: red-and-blues flashing, a cop in a long-sleeved white uniform shirt and navy cap behind the wheel, hand up, gesturing for me to stop.

"Not a fucking chance," I said to Tigre.

I wheeled the Suburban up onto the curb and swung around the cop car, missing the front bumper by a couple of inches. Then I saw that there were more red-and-blue lights, at least five or six municipal police cars up and down the Malecón, blocking the street.

Dirty fucking cops? They knew we were coming?

Tigre betrayed no emotion. I grabbed the steering wheel tighter as I hammered the gas toward the entrance of the hotel, glancing down at the FN Herstal Five-Seven pistol tucked into my waistband.

Could the operation be blown? If so, Bravo will be on the Malecón any second with an army of enforcers, ready for a gunfight. They'll have AKs, hand grenades, RPGs, and all I've got is this Belgian peashooter . . .

I jockeyed the nose of the Suburban in front of the hotel's gate. Surprisingly, the gate was wide open. I saw Brady jumping out of his own Chevy, running hard, disappearing around the back. I knew he was covering the hotel's four-foot wall because he was also worried about Chapo escaping out the back door. Another Cabo San Lucas debacle wasn't going to happen under Brady's and my watch.

Brady grabbed two young marines who were standing near, split them up, and positioned them to watch the wall and the parking garage exit. Once they were in position, Brady went into the lobby just as three marines were grabbing the watchman, searching for hotel room keys behind the desk. Tigre and his men had already made their way inside.

I was standing out front near the pool, with a view of the front of the hotel, my FN Five-Seven aimed at the dark, empty lot to the south as I continued to scan the shadows.

As much as I wanted to be inside smashing doors with Tigre, I knew I needed to make sure our perimeter was tight. I wasn't going to rely on anyone else. Was it completely covered?

Dammit—we need more guys posted back there . . .

Just then, Leroy appeared, walking from the hotel out toward the street.

What the hell is he doing? I said to myself. *He should be inside by now, trying to pin a door.*

Leroy walked out onto the Malecón and pointed up toward the hotel.

He looked at me, then back up toward the front of the hotel.

"Fourth floor," Leroy said. "I'm getting a strong signal on this north end."

Then he gestured with his hands and disappeared quickly back inside the Miramar lobby.

Within minutes, a few lights were flickering on—room by room, floor by floor, the hotel was coming aglow.

Good, we're finally getting somewhere.

I couldn't take it any longer; too much time had passed. If Chapo was planning an escape, he had to be doing it at this very moment.

I began jogging down the hotel ramp to the street—to physically run the perimeter myself, double-checking that enough marines had the sides and back covered—when I heard another loud squelch.

Then Spanish chatter over the radio:

"Ya tenemos el blanco!" I ran up the ramp to Nico, who was holding the radio tight to his ear. "They have the target in custody! They got him!" Nico said.

Another radio squelch:

"Dame un blindado!"

"They need an armored vehicle right now!"

I couldn't hear anything after that echo—*dame un blindado!*— and then there was a piercing silence and I turned, running fast to my Suburban . . .

Pistol in my right hand, I sprinted as fast as I'd ever run anywhere in my life.

I threw the Suburban into drive and gunned it down the ramp into the Miramar's underground parking garage. Three marines were on the ramp, waving me on.

Vamos! Vamos! Vamos!

It was too dark to see anything clearly underground, but knowing that the marines were about to extract Chapo, I immediately repositioned the Suburban, angling the Chevy precisely so it would be ready to exit quickly.

Like clockwork, three more marines emerged, standing up a shirtless man who'd been splayed out on the floor. I could see only a dark silhouette and a brief flash of white skin. He had his hands cuffed behind his back, no blindfold on his face, as they yanked him up off the ground, leading him forward by the silver elevator doors.

The prisoner was short and bare-chested, but I still couldn't make out his face through the thick tinted bulletproof glass of the Suburban; the skin of the man's chest grew increasing pale under the glare of more flashlights.

I jumped out of the driver's seat, still wearing that black hat and balaclava, and ran up to the prisoner.

I stopped abruptly in front of him.

We were face-to-face at last.

I couldn't resist:

"What's up, *Chapo-o-o-o*!?"

How strange it must have been for this drug lord to see someone wearing one of his *own* black hats. Guzmán's eyes bulged, then he hunched one shoulder, flinching, as if he thought he was going to be slugged.

I stared at him, and Chapo held my gaze—just for a moment. There was no mistaking it now: I had my man. That usually fastidious hair—jet black—was greasy, messed up; there was the trademark thick black mustache, and skin so pale it was nearly translucent from all those years of living without daylight, stuck in his rat's world of holes and tubes. Chapo had on black Adidas track pants—they were low-slung, just barely clinging to his hip bones, and they exposed a firm, if Buddha-like, potbelly.

As the marines walked him toward the Suburban, I slapped him on the back—not hard, just an *attaboy!* whack like I once did to my brother, Brandt, after a touchdown, or with Diego after we'd closed some big UC deal.

I pulled my hand back, wet with his sweat. Chapo's back felt like it had been slathered with suntan oil. He probably hadn't showered in days. I hopped into the driver's seat of my Suburban while Chapo was pushed into the rear center seat, flanked by two marines. They'd periodically question him and he'd respond with an almost robotic-sounding *"Está bien—está bien . . ."*

I turned suddenly: *"Mira!"*

Chapo answered me calmly, deferentially:

"Sí, señor?"

I snapped three quick photos on my iPhone.

I spun back around in the driver's seat, transmission in park, foot revving the gas pedal, ready to rock.

Only then did it strike me: we had no exit strategy. The past few weeks had been all about the hunt; we never fully planned for the contingency of having Chapo cuffed and in custody.

Well, I'm going to have to drive this fucker 1,016 kilometers—some twelve straight hours—to Mexico City. Difficult—but doable . . .

But then I got out of the Suburban, knowing that it was far too dangerous for any American agent: a vehicle carrying Chapo Guzmán would be a moving target anywhere in Mexico. One of the SEMAR officers would have to drive.

I turned around, spotting Brady for the first time since we'd arrived at the front of the hotel. We hugged.

"Un-fuckin'-believable!" Brady was shouting, tears welling up in his eyes.

I had never seen him get truly emotional about anything. Brady's usual scowl was now transformed into a broad smile.

I WALKED WITH BRADY from the parking deck up to the street. He was saying something, but I couldn't even hear him; I was still overcome.

We stepped out onto the curb beneath the Miramar sign. The warm ocean breeze swept across my face and slowly began to break my trance. The leaves of the palms whipped in the wind overhead. I turned, bear-hugging Nico, and then Leroy. Both men, along with Leroy's entire marshals team, had been instrumental in the hunt during these final weeks.

I looked up into the dawn sky: the pitch-blackness lightening to a hue of dark blue. I took a long, deep breath, spinning around in the middle of the street, my vision only now coming into full focus.

My family back in Mexico City . . . I hadn't called since that delirious conversation when I was lying flat on my back in Chapo's driveway.

My first text was to my wife.

"Got him, baby."

"No way!"

"Yeah, it's done."

"Coming home?"

"Yes."

"When?"

"Not sure. Very soon."

The darkness was quickly breaking as the sun started to peek over the Sierra Madre to the east. I heard the welcome sound of SEMAR's MI-17 helicopter, far off to the north, a rumble like the thunder of horses' hooves growing ever closer.

QUÉ SIGUE?

BRADY AND I WALKED down the long sidewalk on SEMAR's Mazatlán base toward the interview room. Chino stood there, chest puffed out; he wore a blank expression, and now, for some reason, he was blocking us.

"By order of the secretary of the navy, I can't let anyone in here," Chino said dryly.

"Come on, brother," I said. "After all we've been through?"

"Orders come directly from the secretary of the navy." Chino kept up the dead-eyed stare, then, turning on his boot heel, shut the door.

Brady and I paced outside until the door cracked open.

It was Tigre, gesturing for us to sneak inside.

I saw Chapo seated on a sofa in a clean short-sleeved navy-blue polo shirt. His entire face above the nostrils was wrapped up, mummy-like, in white gauze.

Chapo was talking in a normal tone of voice, no trace of fear or anger, but it was clear that his spirit was deflated. I recognized the voice immediately from my verified recordings; it was a voice I had listened to so many times I'd often dreamed about it.

Now the voice had a strange high pitch to it. Not stress—not exhaustion. Relief, perhaps? The realization that the thirteen-year hunt was, at long last, over?

The interrogation was being conducted with candor and respect. Chino was asking the questions in Spanish.

Chapo began by calmly stating his full name.

"Joaquín Archivaldo Guzmán Loera."

"Date of birth?"

"Fourth of April, 1957."

"Where were you born?"

"La Tuna. El Municipio de Badiraguato, Sinaloa."

I stood back in wonder: I'd written out that full name, date of birth, and town so many times—in my DEA sixes, case updates, PowerPoint presentations—that it had somehow become an extension of myself. I knew it as well as my own Social Security number. Now to hear it all confirmed—in the twangy mountain accent of the squat little man himself—seemed surreal.

Guzmán was no ghost, no myth, no invincible kingpin. He was a captured criminal, like any other, a flesh-and-blood crook, his eyes wrapped in white gauze. He was sitting right there on a sofa, not more than six feet away from me, stating that he had severe tooth pain and recently had had one of his molars fixed.

Chino asked who Guzmán's key operational lieutenant was in the United States.

Chapo paused. "I don't have one," he said finally.

I nodded at Brady; this was backed up by our own intel.

Chino asked him how much weight he was moving from the south. I remember Guzmán saying that his cocaine shipments were between four hundred and eight hundred kilograms at a time. I nodded again. We knew Chapo was being straightforward—gone were the days of the massive multi-ton shipments of blow from South America.

Chino asked how long Chapo had been living in Culiacán.

"Not long. A couple of weeks."

Brady and I looked at each other. That was a bald-faced lie.

Chino said something about the "business" not being what it used to be.

"*Claro que sí*," Chapo said. "There's no respect anymore. I do my own thing. This business now, it's tough. Really tough."

BRADY AND I LEFT the interrogation room, walking out onto the tarmac toward the waiting MI-17 and Black Hawk; we huddled up with Tigre and a group of marines who'd been part of the takedown team, and for the first time I heard the details of what had happened hours earlier that morning on the fourth floor of Hotel Miramar:

When the marines busted through the door of Room 401, Chapo's first line of defense had been Condor. SEMAR quickly apprehended him, then stormed through the two-bedroom suite. In one bedroom they found two women: Chapo's cook, Lucia, and the nanny, Vero, fast asleep with Guzmán's two-year-old twin daughters. The marines raced to the larger bedroom in the back, where they discovered Emma Coronel, Chapo's young wife, who had just awoken.

Chapo had jumped out of bed in his underwear and run into a small bathroom, armed with an assault rifle. As Emma screamed, "Don't kill him! Don't kill him!" Guzmán dropped the gun, offering his empty hands through the bathroom doorway. They took

Chapo down without a single shot being fired and brought him down the service elevator to the parking garage.

Now I watched as Brady helped some of the marines carry Chapo's daughters—still dressed in their yellow-and-pink pajamas—from the Chevy Captiva toward the building in which Chapo was being held.

I walked a little further down the road and saw Condor—we'd identified him as Carlos Manuel Hoo Ramírez—lying in the bed of a pickup truck, handcuffed, his eyes wrapped with gauze like Chapo.

I recognized him as the same man in the photograph we'd found in his house in Culiacán. I pulled out my iPhone and took a shot of the tattoo on his calf: a condor's head. Then I walked down and saw Emma, the cook, and the nanny sitting handcuffed inside another vehicle, their eyes blindfolded, too.

Brady and I continued to hug and congratulate every marine we came across on the base. At some point I realized I was still carrying Tigre's pistol tucked into the front of my BDU pants.

"*Gracias, carnal,*" I said, handing back the black FN Five-Seven to Tigre. I couldn't believe that no one had to fire a shot during the entire operation. Tigre took the pistol and slid it back in his thigh holster.

"*Tu lo hiciste,*" Tigre replied with a grin. "You did it."

Even Admiral Garra managed to crack a small smile when I congratulated him later.

Eventually, Nico, Chino, Chiqui, and several other marines led Chapo, handcuffed, his eyes still wrapped in gauze, out of the interrogation room and placed him inside the Black Hawk. The rotors whipped up clouds of dust and grit; I shielded my eyes with one hand as the helo lifted off the pavement en route to the Mazatlán International Airport, where Chapo was to be flown immediately by Learjet, accompanied by Admiral Furia, to Mexico City, where he would be paraded in front of the world's press.

MOMENTS AFTER CHAPO'S DEPARTURE, Brady and I boarded an MI-17 and took off for a low-altitude flight down the Pacific coastline. Both sets of helo pilots and their crews were SEMAR's best and had been with the brigade since we began in La Paz—Brady and I respected them as core members of our team.

"No such thing as 'crew rest' with these guys," I said to Brady. They were ready to fly their birds anywhere, under any condition, and on a moment's notice.

The pilots swung the MI-17 down low, cruising just above the surface of the ocean, so close that I could see the crests of the waves clearly and felt like I could almost reach my hand out the open window and touch the water. Tourists swimming near the beach were ducking and diving as if the MI were about to strafe them.

After the joyride, we touched down at the Mazatlán airport, from which Chapo had flown out only minutes earlier on the jet.

I knew this was likely the last time I'd see any of these marines. I felt like I was leaving a group of my own brothers—these Mexican warriors had done *everything* to keep all of us American personnel safe.

It was all I'd known for weeks: eating, sleeping, making predawn raids together. Returning to DEA now felt as foreign as the entry to Culiacán had.

I hugged Brady one last time.

And it hit me suddenly—I guess it was a mixture of gratitude

and sadness—that I was losing a partner. I could never have accomplished any of this—I would never have come close to taking
Chapo into custody—without Brady and his entire team of HSI
agents, supervisors, and translators back at the war room in El Paso.

"You tell Joe and Neil they just made history," I said.

"Yes, they did," said Brady.

"Safe travels, brother."

"Whatever happens, we'll always have each other's back—
deal?"

"Deal."

The props began to spin on the DEA's King Air. One of the
pilots called out to me that we were ready to roll. I waved one
last good-bye to the group of marines standing on the runway and
ducked my head as I boarded the plane.

I was alone now in the dark cabin as the King Air ascended
into the sky. I watched out the window as the marines down below
grew smaller, and then the coast of Mazatlán disappeared finally
into the distance.

I WAS MET IN Mexico City on the runway by Regional Director
McAllister, my assistant regional director, and my group supervisor, who congratulated me on a job well done. It was because of the
three of them that I had been allowed to run the investigation—
as I knew how—from beginning to end. They'd given me the lat-
itude and time, and it had resulted in a tremendous win for all of
us at DEA.

And Camila Defusio, the deputy US attorney in Washington,
DC—along with her small team of assistant US attorneys—had
ensured that the judicial process never hindered the operation.

Don Dominguez, back at SOD, and his staff, including the group of intel analysts in Mexico City, had also been behind-the-scenes heroes of the capture.

MY BOSSES DROVE STRAIGHT to my La Condesa apartment, where I walked in and hugged my wife and sons.

I wiped a tear from my cheek, thankful to be home, then sat down at the kitchen table. The rustic SEMAR meals had been satisfying, but they were nothing like a home-cooked dinner prepared by my wife. We barely said a word at the table as we ate slowly. We were just grateful to be together again.

The next morning, I went with my wife and sons on a bike ride through the city, just as we usually did on the weekend. Paseo de la Reforma was closed to cars on Sundays and was swarming with bicyclists, runners, walkers, and rollerbladers.

On the newspaper stands, every paper—*Reforma, Excélsior, El Universal, Milenio*—had Chapo's face plastered on the front and banner headlines.

CAPTURAN A EL CHAPO!
CAYÓ!
AGARRAN A EL CHAPO!
POR FIN!
CAE EL CHAPO!

To stand there on Paseo de la Reforma, buying those papers like a local, after all the weeks I'd spent embedded with SEMAR, was like living in another life. I bought all of them and stuffed them in the front basket of my Raleigh. I was struck by the eerie sense of being covert again, chameleon-like, blending in with the bustling crowd. No one snatching up copies of the newspapers could have suspected that the blond-bearded cyclist in a V-neck cotton shirt, shorts, and *chanclas* had been at the heart of the hunt, that only hours earlier I was the agent who led the capture of the most wanted criminal in the world.

THE NEXT MORNING, I put on my suit, knotted my tie—like on any typical Monday—and drove to the embassy in my armored Tahoe.

But as I walked the halls, I felt like a zombie: my body was present, but my mind was not. I walked back to my desk and heard another agent talking about pinging phones of targets in his own trafficking investigation. I felt unsteady, and the office seemed to rock. I felt a sudden blood-pressure drop, and a queasiness, as if I were going to throw up right there on my desk.

I'd been expecting to feel euphoric after the Chapo capture, but I felt the opposite. Over the next few days, I tried to shake it off, but the void only deepened.

SINCE HIS CAPTURE in February, Chapo had been interviewed at the Mexican Attorney General's Office (PGR) before being locked up in the country's most secure prison, Federal Social Readaptation Center No. 1 (Altiplano), in central Mexico, not far from Toluca.

I later heard a story about a remarkable exchange between PGR lawyers and Guzmán. Interrogators apparently said they could now close out the estimated thirteen thousand homicides credited to Chapo.

"*Thirteen* thousand?" I was told Chapo responded, seeming

genuinely surprised. "No, not thirteen thousand. Maybe *a couple* thousand . . ."

Whatever the body count, Guzmán was supposedly now no longer a threat: the authorities assured the public that he was under twenty-four-hour video surveillance at Altiplano. The maximum-security prison housed Mexico's most violent and notorious narco-traffickers and was considered escape-proof.

Guzmán was behind bars, but there was still more bloodshed back on his home turf. On April 10, 2014, I picked up one of the local papers and read that the body of Manuel Alejandro Aponte Gómez—Bravo—had been found dumped on a dirt road near La Cruz de Elota, Sinaloa. Bravo had reportedly been tortured before being shot several times, and had been killed along with two of his associates. No one knew for sure, but street rumor quickly spread that Bravo had died for the unpardonable mistake of not properly protecting his boss while he was on the run in Mazatlán.

SEVERAL DAYS LATER, after the murder of Bravo, I flew with Tom McAllister and my group supervisor to Washington, DC. There I briefed DEA Administrator Michele Leonhart and her top brass at headquarters in Arlington, Virginia. The briefing room was standing room only as, slide by slide, I walked through the details of the operation. Leonhart concluded by congratulating me and our entire team on the capture of Chapo.

"I've been with DEA a long time," said the special agent in charge of SOD, "and this has got to be the best case I've seen in my career."

Immediately afterwards, I jumped in a black armored Suburban with Administrator Leonhart and rode in her motorcade across the

Potomac River to the United States Department of Justice building to brief US Attorney General Eric Holder.

"This was Bobby Kennedy's office when he was AG," one of Holder's aides told me as we entered. I glanced up at the painting of Kennedy, wearing a bomber jacket, hanging on the wall alongside other former attorneys general.

I shook AG Holder's hand and could immediately sense his sincerity and his genuine interest about the details of the operation. McAllister took him through my presentation while I highlighted the story with details from the weeks on the ground in Sinaloa—from our discovery of the Duck Dynasty hideaway through the days of smashing down doors in Culiacán, right up to the predawn face-to-face capture at the Hotel Miramar.

The attorney general asked about Chapo's first escape through the bathtub tunnels.

"Well, we knew he had a tunnel, sir," I said. "But not under *every* house."

"How many houses were there?"

"He had five safe houses in Culiacán," I said. "And they were all connected through the sewers."

Holder was impressed by the persistent hunt and inquired about how we'd been able to sustain our operations until the very end.

"We used Chapo's homes as bases," I explained. "We essentially turned them into makeshift barracks, all of us living on top of each other. We cooked in his kitchen. Slept in his beds."

At the end of the thirty-minute briefing, Attorney General Holder expressed his official thanks on behalf of the Obama administration and the American people for bringing Guzmán to justice. He said this would go down as one of the greatest achievements of the administration.

"So what do you do now?" Holder asked.

I stared at him—not fully understanding—and then Holder added:

"Seriously, what do you do *next*? Take some time off and drink mai tais on the beach?"

Everyone in the room laughed.

"That's what I'm still trying to figure out, sir," I said.

WHAT *COULD* I DO NEXT? The AG's question kept resonating in my mind as I returned to Mexico City. Back at the embassy, I still felt that aching hollowness—and it wasn't subsiding.

Qué sigue?

I'd achieved the greatest challenge possible as a drug enforcement agent, and I realized there was nothing left for me to do at DEA. I had nothing left to give. I couldn't go back to tracking down some lesser trafficker—pinging phones, debriefing sources, gathering intel, crunching numbers—here in Mexico, or in any other country, for that matter.

In my world—among all of DEA's international targets—who was *bigger* than Chapo Guzmán?

In fact, the past few years had *never* been about Chapo; they had only been about the hunt, and now the hunt was over.

I ALSO HAD TO consider the risk to my wife and young sons, exposed as we all were in the heart of Mexico. No one had assigned us any added security or made plans for us to catch the next plane out of the country.

That came with the gig: you take down a drug trafficker—even one as infamous as Chapo Guzmán—and it's back to business as usual.

But no matter how hard I tried, I couldn't stop hearing that phrase echoing constantly in my ears:

"Everything is fine in Mexico until suddenly it's not."

With the security concern and the strong desire to pursue *another* challenge—my next hunt—less than nine months after the capture, I resigned from the DEA, boarded a flight with my wife and sons, and disappeared as fast as Chapo had escaped from me in Culiacán.

EPILOGUE: SHADOWS

AT ALTIPLANO PRISON, on Saturday, July 11, 2015, at exactly 8:52 p.m., Chapo Guzmán could be seen on the overhead surveillance video taking a seat on his narrow bed, changing shoes, then quickly ducking into the shower stall in the corner of his cell. He disappeared behind the low wall separating the shower from his cell, the only spot in the five-by-six-foot prison cell hidden from cameras.

Then he vanished from view and was gone, disappearing into a twenty-inch-square opening that had been cut into the floor. He squeezed into a narrow vertical shaft to the tunnel below, climbed down a ladder, and entered a sophisticated tunnel nearly a mile long. Electric lights had been hung from the ceiling, as had PVC pipe, which pumped fresh air the length of the passageway. Chapo Guzmán's latest tunnel had A/C.

Metal tracks had also been laid the entire length of the tunnel so that an ingenious getaway vehicle—a railcar rigged to the frame of a small modified motorcycle—could be driven rapidly by the escapee. The walls were only about thirty inches apart, jagged and unshored—and barely wide enough for Chapo's shoulders.

The tunnel began beneath a ramshackle cinder-block house, still under construction, in the nearby town of Santa Juana. By the

time the prison alarm sounded and a massive search began, Chapo
Guzmán was once again in the wind.

THE AUDACITY OF CHAPO'S escape plan—to have his chief tunnelers,
most likely Kava and his crew, dig right up under the most secure prison in Mexico—shocked the world. But the method was
certainly no mystery to me or anyone else who'd studied Chapo
for years. As with Chapo's breakout from Puente Grande back in
2001, the escape came with that other Guzmán hallmark: layers of
corruption and bribery.

The official Mexican version of events was quickly dismissed
as a farce. Reports of loud drilling in concrete had gone unheeded;
the supposed blind spot on the video surveillance turned out to be
merely a case of prison staff selectively ignoring the activities in his
cell.

In the moments before his escape, Chapo—appearing fidgety
and anxious—repeatedly goes over to the shower area to check on
activity behind the short wall, and even bends down, apparently to
help pry something open. The video also appears to show an iPad
lying near Guzmán's bed, despite the fact that cell phones, tablets,
and other electronic devices are specifically banned in the prison.

According to a review conducted months later by the Mexican Congressional Bicameral Commission on National Security,
Chapo had never been treated like a typical inmate at Altiplano. In
the seventeen months he spent there, he'd been granted extraordi-
nary privileges, receiving 272 visits from his lawyers alone, as well as
18 family visits and 46 conjugal visits. Perhaps the most sensational
of these latter visits was a reported New Year's Eve rendezvous with
a local Sinaloa politician, a young female deputy from the National

Action Party named Lucero Sánchez López, who was accused of sneaking into the prison with false documents and spending the night with Guzmán. Sánchez forcefully denied these charges, but was nonetheless stripped of her parliamentary immunity.

On June 21, 2017, Sánchez was arrested by US federal agents at the Otay Mesa Cross Border Xpress—the bridge connecting Tijuana's A. L. Rodríguez International Airport with San Diego—and the next day the former legislator was charged in California federal court with conspiracy to distribute cocaine. After the capture, and reviewing information from the line sheets, Brady and I suspected that Sánchez was the same "girlfriend" who Picudo had told us escaped through the tunnel and sewer with Chapo in Culiacán just before we arrived at his safe house.[*]

Upon his escape, Guzmán again catapulted back to the status of world's most wanted fugitive. Interpol issued a "Red Notice" for his immediate arrest. There were sightings of Chapo reported via Twitter, of him supposedly enjoying himself at an outdoor café in Costa Rica. Some rumors were laughably far-fetched: Guzmán was reported to have traveled as far south as Patagonia, Argentina, where witnesses claimed to have seen him in a "sweet shop"—police and military units were on high alert that he was traveling in the Andes, on the verge of crossing the border into Chile.

In truth, Chapo had never left the comfort zone of his own mountain home.

---

[*] The complaint filed against Sánchez alleges that she has continued to deny reports that she was Guzmán's lover. However, in a probable-cause affidavit attached to the federal criminal complaint, a cartel member cooperating with US investigators alleged that Sánchez was indeed Guzmán's long-standing girlfriend, and that she'd admitted to having fled with Chapo through the tunnel on February 17, 2014, moments before SEMAR entered the drug lord's safe house in Culiacán.

———

FROM THE MOMENT the news broke of Chapo's escape, an intense manhunt commenced. Admiral Furia and his Mexico City–based SEMAR brigade took the lead once again, using our operational blueprint and years' worth of intelligence as their guide. SEMAR, working in conjunction with PGR and the Mexican Federal Police, arrested Araña, Chapo's most trusted pilot, who was suspected of flying the kingpin up to the Sierra Madre of Sinaloa immediately after his escape from Altiplano.

Guzmán, no doubt feeling more untouchable than ever after the brazen breakout, didn't even bother switching up his telecom-munications system. He may no longer have had Condor to act as his faithful secretary, but Mexican authorities were able to inter-cept BlackBerry PIN messages of Chapo's closest associates—just as we had done for months.

During his time hiding out in the mountains, Kate del Castillo—the star of Guzmán's favorite telenovela, *La Reina del Sur*—resurfaced and was communicating with Chapo through various BlackBerry mirror devices. Even as a fugitive, Chapo was still seeking to have his life story told on the big screen—just as he had done with Alex Cifuentes back in October 2013. Chapo was also still clearly infatuated with Kate, so thrilled to meet her that he almost disregarded those who planned to come with her, including actor Sean Penn—Chapo had never even heard of the Hollywood star—but Kate assured him that Penn could facilitate the production of Chapo's movie.

Chapo's narcissism unwittingly led him into a trap—a varia-tion of the *Argo*-style operation Brady and I had strategized two years earlier. On October 2, 2015, Guzmán agreed to a face-to face meeting with Castillo, Penn, and several others at a secluded location high in the Sierra Madre, along the Sinaloa-Durango border. As reported in the Mexican media, Mexican authorities already had Castillo, Penn, and Chapo's lawyers under surveillance the entire time. The meeting was reportedly a tequila-fueled dinner and sit-down with Kate's Hollywood friends to develop his life story. Sean Penn, it turned out, was playing the role of journalist, on assignment from *Rolling Stone* to write an exclusive article. When it was published later ("El Chapo Speaks"), Guzmán said little of note. The meandering ten-thousand-word, first-person account was widely derided as self-indulgent and naive, and it got particular heat for the arrangement *Rolling Stone* had agreed to in which Guzmán—or more likely his attorneys—got approval over the final copy.

According to Castillo, after dinner Guzmán had departed abruptly; he'd said it wasn't safe for him to stay overnight at the same location as his guests. Several days later, SEMAR conducted helicopter raids in some of the mountain villages outside Tamazula, Durango, but were caught in a hail of gunfire from Chapo's security men on the ground. Once SEMAR finally made entry into one of the homes near Tamazula, they discovered BlackBerrys, medications, and two-way radios. Once again, Chapo had escaped by mere moments out the back, down a steep hill and into a ravine, and was reported to have injured his face and leg.

With SEMAR forces closing in from the south, making dozens of raids in the tiny mountainous villages, where Chapo could typically hide without worry, he had no choice but to flee north through Sinaloa.

His network of safe houses in Culiacán was obviously no longer an option. And with Bravo dead, Chapo drove directly into the

hands of the only chief enforcer left on his payroll, the feared
Cholo Iván, up in Los Mochis. SEMAR continued to track Chapo
the entire time, as he settled in on the Pacific coast, taking refuge
in a comfortable safe house constructed on a design similar to the
ones in Culiacán. .

IN THE RAIN AND DARKNESS on Friday morning, January 8, 2016,
SEMAR launched Operación Cisne Negro ("Black Swan"). Units
of masked marines approached in *rápidas* with their headlights off,
military helicopters hovering overhead, surrounding a white two-
story house in a middle-class neighborhood of Los Mochis, where
they'd confirmed that Chapo and Cholo Iván were hiding.

Around 4:30 a.m., SEMAR began their entry into the house
through the front door and were met by immediate gunfire. The
marines advanced slowly while tossing grenades and laying down
heavy fire from their assault rifles. After more than twenty minutes
of fighting, five of Chapo's gunmen lay dead, six were injured, and
several would be arrested. Only one marine was wounded in the
firefight.

But with the time it had taken SEMAR to gain access to the
safe house, Chapo and Cholo Iván were long gone. A search of the
house revealed two tunnels: one beneath the refrigerator, the other
in a bedroom closet. A switch near a lightbulb activated a trapdoor
behind the mirror, opening to an escape ladder and a passageway
directly into the sewers of Los Mochis. It was Chapo's signature MO.

Once Chapo and Cholo Iván had reached the sewer—only one
meter high, flooded due to the heavy overnight rains—they had
to crawl slowly for blocks on their bellies through fetid water and
human waste.

Less than an hour later, Chapo and Cholo Iván emerged from the sewer. The two fugitives forced open a square metal manhole but had trouble lifting the hinged cover, so they wedged in one of their shoes to prop it open. In the sewer, they left behind an AR-15 equipped with a grenade launcher.

Chapo's luck was running out. According to media reports, Chapo and Cholo Iván brandished their guns and carjacked a white Volkswagen Jetta after they emerged from the manhole in the street. But amazingly, the Jetta quickly broke down, and after driving only a few blocks, Chapo and Cholo Iván ditched the Volkswagen. At a traffic light, they carjacked a red Ford Focus, reportedly driven by a woman with her daughter and five-year-old grandson.

Six miles before they reached the town of Che Ríos, the Ford Focus was stopped by Federal Police. Cholo Iván exited the vehicle armed with a weapon, while Chapo was crouched in the backseat.

The media also reported that Chapo offered to reward the police with homes and businesses in Mexico and the United States and promised them to "forget about working for the rest of their lives." All they had to do was let him go. The cops refused the bribes and put Chapo and Cholo Iván into a patrol car.

The cops also snapped a photo and sent it to their superiors. It showed Chapo sitting in the back of the cop car, wearing a filthy tank top, next to shirtless, grim-faced Cholo Iván.

Authorities feared the arrival of more gunmen. To avoid a shoot-out, they drove Chapo and Cholo Iván to Hotel Doux, just outside Los Mochis, where they holed up in Room 51 until additional Federal Police and SEMAR arrived.

Chapo and Cholo Iván were then flown to Mexico City; Guzmán found himself back in Altiplano, the same maximum-security prison he'd tunneled out of the previous summer. Chapo's six months on the run—six months of embarrassment for the government of Mexico—had finally come to an end.

"Mission accomplished," President Enrique Peña Nieto announced on his Twitter account. "We've got him."

HOW COULD MEXICO POSSIBLY ensure that Chapo wouldn't attempt yet another escape from custody? Prison officials announced that security at Altiplano had been revamped for Guzmán's arrival. They cited the installation of hundreds of new surveillance cameras, motion sensors in air ducts and underground, and reinforced steel concrete floors. They also deployed dogs trained specifically to detect Chapo's distinct odor and would constantly move him between cells, followed closely by a team of guards.

Then, in the early-morning hours of Friday May 6, 2016, Chapo was transferred, without warning, to a prison outside Ciudad Juárez, reportedly due to its proximity to the border and to facilitate a rapid extradition to the United States. Chapo soon was complaining about the inhumane and unbearable conditions; his Juárez prison cell was so filthy that he'd asked for bleach to clean it himself. According to his lawyers and the report of the psychiatrist who visited him, the kingpin was badly deteriorating: he was "depressed and suffering hallucinations and memory loss because of harsh conditions in the prison where he is jailed."

Chapo told the doctor that "psychological torture" was being inflicted on him. Lights in his cell were kept on twenty-four hours a day, and his only human contact was with masked corrections officers. He also reported that he was being woken up every four hours to appear on camera for an inmate roll call. "They do not let me sleep," Chapo said, according to the psychiatrist's report. "Everything has become hell." Guzmán claimed to be taking a cocktail of thirteen pills daily—for pain, insomnia, and constipation. His sleep deprivation and hallucinations were so severe that he felt he was on the verge of death. "They have not beaten me," Chapo said. "But I would prefer that."

On October 24, 2016, Emma Coronel filed an official grievance with the National Human Rights Commission, alleging that the new prison conditions were inflicting "irreparable" psychological damage on her husband. She claimed that being confined in the Juárez prison would either kill Chapo or make him "go crazy" in a matter of months. She also complained that her conjugal visits with her husband had been reduced from four hours a week to only two hours.

Mexican officials denied that Guzmán's rights were being violated—he was being treated as a high-profile prisoner who'd made two previous prison escapes—and suggested that the reports of mistreatment were merely a legal strategy on the part of the cagey drug lord.

AND WHAT OF CHAPO'S status as arguably the most powerful narco-trafficker of all time? The truth was, Chapo's hold on his sprawling narcotics operations back in Sinaloa was beginning to crumble.

His most trusted sons—Iván, Alfredo, Güero, and Ratón—remained at large, but they did not command a modicum of the respect accorded their father. Many integral members of his inner circle were either dead—like Bravo—or in custody, like Condor, Cholo Iván, Picudo, and Araña.*

Even Chapo's mother was no longer seen as untouchable. In mid-June 2016, it was reported that some 150 gunmen stormed into Guzmán's hometown of La Tuna, killing three people in the community and even looting Chapo's mother's home, stealing several vehicles. Eighty-six-year-old Consuelo Loera de Guzmán wasn't harmed, but the ransacking of her son's childhood home, the mountain hacienda at which Chapo had often taken refuge, was viewed as incontrovertible proof throughout the narco world that Guzmán no longer had power over his cartel.

Chapo was facing numerous legal cases in Mexico, primarily for drug trafficking and murder, but the government indicated they no longer had interest in prosecuting him at home; in early 2016, President Peña Nieto announced that he'd directed his attorney general's office to "make the extradition of this highly dangerous criminal happen as soon as possible."

Guzmán faced US federal prosecution for alleged involvement in cocaine, marijuana, and heroin trafficking, racketeering, money laundering, kidnapping, and conspiracy to commit murder. Jurisdictions in Arizona, California, Texas, Illinois, New York, Florida, and New Hampshire all staked claims to prosecute him on various offenses related to his status as boss of the Sinaloa Cartel.

Most legal experts agreed that, once extradited, Chapo would likely be sent to the Eastern District of New York—the Brooklyn venue where infamous Mafia bosses like John Gotti stood trial in the 1980s and '90s.

---

* We never did confirm the identity of Lic-F, even if I still held my suspicions. In early May 2017, the Mexican attorney general's office announced that authorities had arrested Dámaso López at an upscale apartment building near downtown Mexico City.

Loretta Lynch, then US attorney for the Eastern District—later United States attorney general—had personally signed the indictment, filed on September 25, 2014, charging Guzmán and other alleged members of his cartel with conspiring to import tonnage of cocaine into the United States between 1990 and 2005.

The indictments allege that Guzmán employed sicarios to carry out hundreds of acts of violence in Mexico, including murder, torture, and kidnapping. Lynch called Chapo's Sinaloa Cartel "the largest drug trafficking organization in the world," responsible for the vast majority of drugs imported into the United States.

YET, GIVEN CHAPO'S REPUTATION as the king of modern-day escape artists, it was perhaps inevitable that in July 2016, Internet rumors claimed that Guzmán had escaped from the lockup in Ciudad Juárez.

The response of the Mexican government was instantaneous: Secretary of the Interior Miguel Ángel Osorio Chong released a photo on his Twitter account showing Chapo sitting alone in the brightly lit and desolate prison room, clean-shaven, surrounded by only a few hidden, shadowy guards, waiting out the clock before his extradition to face justice in the United States. *"Para los rumores, una imagen,"* Osorio Chong wrote. "For the rumors, an image . . ."

IT SEEMED AS IF Chapo's legal team would drag out the judicial process for many months, but then, on January 19, 2017—and without warning—the Mexican Foreign Ministry and the US Department of Justice abruptly announced that Guzmán was being extradited.

Chapo was transferred from the prison handcuffed and still

in his gray jailhouse jumpsuit, wearing an oversize tan jacket, his face pale and his hair so closely cropped that he looked like a skinhead. Chapo was clearly agitated and frightened as he sat aboard the Mexican government's Challenger 605 jet, which departed for New York just after 5:30 p.m. Several hours later, the plane landed at MacArthur Airport, in Islip, Long Island; Chapo was taken into US custody and escorted off the plane by agents from the DEA and HSI.

The timing of the extradition seemed highly unusual, and the government of Peña Nieto offered no explanation for why it chose to send its most notorious prisoner to the United States on the last night of President Obama's term in office.

From Long Island, Guzmán was taken to his new temporary home in the heart of lower Manhattan, the Metropolitan Correctional Center, a blocky beige twelve-story complex wedged between the Brooklyn Bridge and the Manhattan Bridge on Park Row. One of the country's most secure federal lockups, this is the prison where other high-profile inmates have awaited their trials, among them the Gambino crime family boss John Gotti and terror suspects such as the Al Qaeda associates of Osama bin Laden and Ramzi Yousef, the mastermind of the 1993 bombing of the World Trade Center.

Guzmán was housed in the most high-security wing within the MCC—10 South—known as "Little Gitmo."

On January 20—while most of the world watched President Trump's inauguration in Washington—Guzmán was brought before a judge in the Eastern District, in downtown Brooklyn, where he heard the seventeen-count indictment, alleging that between 1989 and 2014, as the leader of the Sinaloa drug cartel, he'd run a "criminal enterprise responsible for importing into the United States and distributing massive amounts of illegal narcotics and for conspiring to murder people who posed a threat to the narcotics enterprise." The United States government demanded that Chapo surrender $14 billion "in drug proceeds and illicit profits" that he allegedly smuggled into Mexico from the United States.

"Today marks a milestone in our pursuit of Chapo Guzmán," said Robert Capers, the US attorney for the Eastern District. "Guzmán's story is not one of a do-gooder or a Robin Hood or even one of a famous escape artist. [His] destructive and murderous rise as an international narcotics trafficker is akin to a small cancerous tumor that metastasized and grew into a full-blown scourge that for decades littered the streets of Mexico with the casualties of violent drug wars over turf."

The US prosecutors claimed that Chapo had continued to run his narco empire even during his incarceration in Mexico's prison system. "He's a man known for a life of crime, violence, death, and destruction, and now he'll have to answer for that," Capers said.

IT DIDN'T TAKE CHAPO LONG to begin complaining about the harsh conditions in Little Gitmo. On February 3, 2017, a heavy police presence escorted Guzmán from the MCC to the Brooklyn federal courthouse. The scene was unprecedented in New York history, even by the standards of the city's biggest criminal trials. Not even notorious mobsters like Gotti or any high-profile terrorists had been transported under such heavy security.

A twelve-car caravan, with Guzmán hidden behind the heavy tint of an armored van, shut down the outbound Brooklyn Bridge for fifteen minutes during the height of New York's morning rush hour. Authorities said they were concerned that Chapo had the resources to launch a possible "military-scale" rescue.

Guzmán appeared before the judge wearing a navy prison uniform, turning at one point to smile at his wife, Emma, who was seated in the first row of the courtroom. It was the first time they'd seen each other since the extradition.

Chapo's defense attorneys sought to lighten some of the strict security measures at the MCC. They complained that Guzmán was on a twenty-three-hour lockdown in jail, allowed to leave his cell only to speak to limited members of the defense team and granted one hour of exercise a day. One of his court-appointed attorneys, Michelle Gelernt, called the security measures "extremely restrictive" and said that Chapo should at least be allowed to make phone calls to his attorneys and have face-to-face visits with his wife.

Chapo's lawyer argued that he had caused no security problems since he had come to the United States, and that the current restrictions were excessive.

But Judge Brian Cogan—without mentioning Chapo's two prior prison escapes—was clearly unimpressed. Regarding the extra security measures, the judge deadpanned, "We know the reason for that."

IT WAS A BREEZY midsummer Saturday evening. I was back in Arizona for a friend's wedding, and Diego picked me up at Sky Harbor International Airport in his Chevrolet Silverado.

It wasn't the old Black Bomber, but the booming stereo system quickly brought back memories of our time together—ten years earlier—on the Phoenix Task Force.

"Paraíso personal de la dinastia Guzmán entre bungalows y alberca,"

Diego was singing as he hit the gas onto the freeway. *"Lo querían asegurar al más grande de los grandes—Señor Chapo Guzmán."*

> *A personal paradise for the Guzmán dynasty*
> *Among the palapas and the pool*
> *They wanted to capture the biggest of them all*
> *Señor Chapo Guzmán*

It sure felt like old times as we shot west to the Maryvale neighborhood once again for *mariscos* and a few beers.

We drove into a nearly blinding sunset, the Phoenix Mountains and the towering saguaro cactuses welcoming me like old friends. Over his iPhone's Bluetooth connection, through the Bose speakers, Diego was blasting the narcocorrido "La Captura del Chapo Guzmán," by Jorge Santa Cruz. He sang the lyrics loudly—syllable for syllable—as he drove the Silverado west on Interstate 10.

I no longer needed help with the Spanish translation. And I remembered what Diego had told me in my early years at DEA.

You're no one in the narco world until you've got your own corrido.

I was impressed by the accuracy of "La Captura del Chapo Guzmán." Virtually every detail of the operation was covered in the verses of the song: the arrival of the marines to storm Chapo's personal paradise at Duck Dynasty; our plan B to catch Chapo unaware in his house on Río Humaya in Culiacán; Chapo's Houdini-like disappearance into a secret passage beneath the bathtub and into the city's drainage system; Picudo's confession that he'd dropped off *"el más grande de los capos"*—the biggest of bosses—on the road to Mazatlán, where El Bravo was there to protect him.

And, ultimately, how the marines closed in, in the predawn hours of February 22, for the surprise raid on Room 401.

> *A Mazatlán, Sinaloa*
> *Un lugar paradisiaco*
> *Elementos de Marina*
> *Uno a uno fue llegando*
> *Pa no levantar sospechas*
> *En el Hotel Miramar*
> *El 22 de Febrero*
> *Cayó El Chapito Guzmán*

"Man, this is all you," Diego said. "The fall of Chapito. Your own corrido, Drew. *Felicidades!*" He let out a loud laugh. "You made it, brother."

I nodded as Diego pulled off the 10 at exit 138, turning his Chevy onto 59th Avenue.

But as faithful as the corrido was to the details of the capture, there was no mention in the lyrics about Americans having boots on the ground, no shout-out to DEA—those three letters most feared by every narcotrafficker.

"Yeah, man," I said, "but it's missing something."

"What's that?"

"There's nothing about Las Tres Letras."

"True," Diego said. "They didn't know about Las Tres Letras."

"Como siempre," I said, smiling. "In the shadows. Like always."

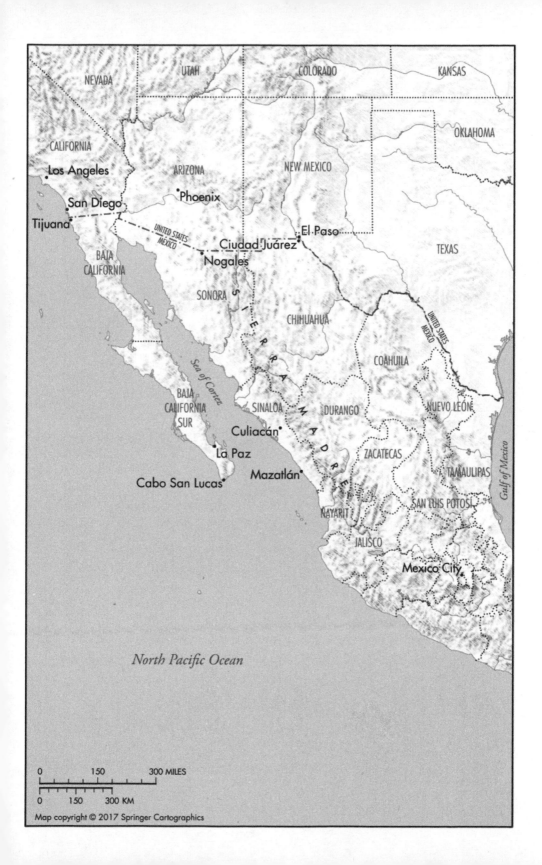

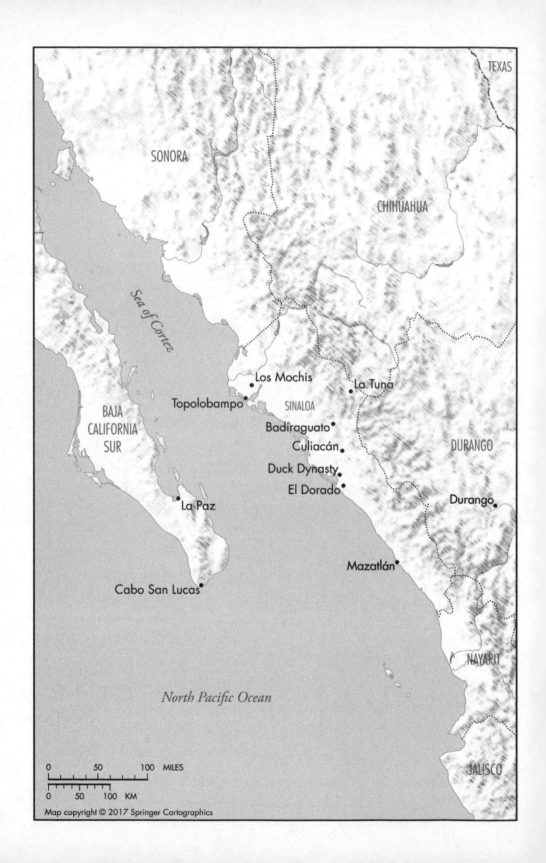

TEXAS

SONORA

CHIHUAHUA

Sea of Cortez

BAJA
CALIFORNIA
SUR

Los Mochis

Topolobampo

La Tuna

SINALOA

Badiraguato

Culiacán

DURANGO

Duck Dynasty

El Dorado

La Paz

Durango

Mazatlán

Cabo San Lucas

NAYARIT

North Pacific Ocean

JALISCO

0   50   100 MILES

0   50   100 KM

ACKNOWLEDGMENTS

THE CAPTURE OF the world's most-wanted drug kingpin could never have been achieved by one man alone. During my law enforcement career, I had the pleasure to work with hundreds of men and women who deserve my deepest thanks here, but security concerns prevent me from directly naming them.

First and foremost, I owe immense gratitude to my wife for her unwavering love, support, and patience. The time she sacrificed over the years has been an essential contribution to my success, and I can only hope my continued love for her offers some solace for my constant preoccupations. My sons—the greatest accomplishments of my life—have put the true meaning of this story into focus. They will always be my inspiration.

Thanks also to my parents, grandparents, in-laws, siblings, cousins, aunts, uncles, nieces, nephews, and all my other family and friends who followed my journey so closely and were a constant source of love and support through it all. Special thanks go out to two very good friends, one who pushed me toward a career with DEA, back when I first had no intention of pursuing one, and a second who's been there at all the right times, graciously creating opportunities for my family and me to thrive after my days in law enforcement.

Without Diego Contreras, my career at DEA would never have evolved as quickly as it did. We grew together, complementing each other's strengths—becoming a powerful team. I will forever be grateful for our partnership. Diego's cunning and savvy approach to investigation is remarkable; he was the initial driving force that ultimately led to the capture. His innate ability to infiltrate Mexican drug-trafficking organizations will mark him as one of the greatest undercover investigators in the history of federal law enforcement. Though miles may separate us, I will always consider him a brother and a partner for life.

The Drug Enforcement Administration is one of the premier law enforcement agencies in the world, and I'm honored and fortunate to have served alongside some of DEA's finest. Special thanks to former DEA administrator Michele Leonhart and her former staff at DEA Headquarters; Regional Director Tom McAllister and his entire staff, including my assistant regional directors and group supervisor, as well as my fellow special agents, pilots, intelligence analysts, and administrative support personnel. Special thanks to Nico Gutierrez for all his assistance with translating and for his frontline coordination.

This mission would never have been accomplished without the former SAC of the Special Operations Division and his staff, especially Don Dominguez and his team, my former ASAC and two Group Supervisors at NTF, as well as all my previous colleagues in Team 3. Finally, a very special thank-you to Snake for greasing all the right skids.

When I first ran into Homeland Security Investigations Special Agent Brady Fallon, it felt as if we'd known each other for years. His humility, good humor, and drive served as a catalyst for creating and sustaining HSI and DEA's relationship, which ultimately led to our success. No one will ever quite understand what it took to lead an investigation and operation of this complexity and magnitude—but Brady certainly knows, because we did virtually every step of it together. I owe Brady, Neil Miller, and Joe Dawson a world of gratitude; they were the workhorses behind the operation—and truly its unsung heroes. And I'm indebted to the tenacious team of HSI senior executive staff, supervisors, special agents, intelligence analysts, and translators; without their diligent work and coordination, the capture would never have happened.

United States Marshal Leroy Johnson and his team deserve far more credit than the space constraints of the story allow, and are rarely afforded the accolades they deserve for their bravery. Once we had boots on the ground in Sinaloa, it was the marshals' technical and operational expertise that proved essential. I give my sincere thanks to each and every one of them.

There are so many people within the US Department of Justice and US Attorney's Offices throughout the nation who deserve my gratitude—but none has been as instrumental to this operation as then–deputy chief Camila Defusio and her team of assistant US attorneys. Their relentless efforts provided us the perfect tools to track down and capture the most elusive drug lord of our age. I also owe many thanks to several former AUSAs from San Diego and Chicago with whom I had the pleasure to work closely over the years; their support had a defining impact on the investigation leading to the capture. We couldn't have done it without all of them.

At the Lincoln County Sheriff's Office, I'd like to thank all those—past and present—for opening the door to my law enforcement career and offering continued support in the years after I left.

Having the privilege to live and work in a country I've come to love almost as much as my own can be attributed to only one factor: the people of Mexico. Men and women who exude great pride in their land, who open their homes, share their culture, and believe in the common good. SEMAR admiral Furia and his marines are these types of Mexican patriots. *Todo por la patria.* All for the homeland. These Mexican marines live—and many good men have died—for those words. Admiral Furia and his brigade understood the importance of trust and embraced our partnership wholeheartedly.

Bonds between nations—even those as vast as the United States and Mexico—often boil down to just a handful of personal relationships. And I know of no better example of how both countries operated together as one team and achieved what most thought was impossible. It's been an incredible honor to work so closely with SEMAR; I'll never be able to pay back the debt of gratitude I owe to Admiral Furia and every single marine involved in this operation. They protected our lives—ensuring we returned home safely to our families without *once* having to fire a shot. They are warriors in the truest sense of the word, and I'll forever cherish their brotherhood.

A great deal of thanks go to the Mexican Federal Police, the Mexico Attorney General's Office (PGR), and all the specialized narcotics units throughout Latin America with whom Diego and I worked so closely in the early years. These men and women must fight against systemic institutional corruption every day—yet somehow manage to succeed and improve the quality of life for their citizens by disrupting some of most violent international drug-trafficking organizations. I'm honored to call many of them friends.

Thanks also to the members of Canadian law enforcement—

specifically, the Vancouver Police Department and the Quebec Provincial Police—for their efforts and support of DEA's mission.

I owe Douglas Century, my coauthor, an immense amount of gratitude. His complete submersion and selfless dedication were critical in capturing every detail, fact, and feeling of my journey. It has taken us several grueling years to fine-tune every line, paragraph, and page. Quite simply, I could never have written this book without him.

My agency, 3Arts Entertainment, was instrumental in bringing us together and helping us conceptualize how best to render this story; I owe everyone at 3Arts and the entire team at HarperCollins my deepest gratitude.

When I was in the DEA Academy, just before we graduated, we created a class T-shirt which read:

> Woe to the wicked! Disaster is upon them!
> —Isaiah 3:11

It wasn't that we were so religious, but we all felt that sentiment deeply: No matter how big or small the criminals, how distant or secretive the lair, there will always be lawmen and lawwomen dedicated to bringing them to justice. No criminal can operate with impunity forever.

Woe to the wicked! It's a phrase that has resonated with me since that academy graduation. So, my final note of gratitude goes out to all the heroes of our nation's law enforcement and military who, every day and night, commit their lives to delivering "disaster to the wicked" so that we can *all* sleep in peace.

—A.H.

A NOTE ON SOURCES

WRITING A BOOK set in the murderous *milieu* of contemporary narco-trafficking can be daunting. As with any criminal underworld, what passes for official history is often mere speculation or mythology. It's nearly impossible to separate fact from fable: urban legends, prison lore, and old war stories get repeated generation after generation—reprinted in newspapers, magazines, blogs, and books—to the point that they're often indistinguishable from verifiable fact.

It's no less true for the early days of Joaquín Guzmán than it was for American gangsters such as John Dillinger or Pretty Boy Floyd, Al Capone, or Bugsy Siegel.

The United States and Latin America are rife with narco-*porn* today—salacious films, paperbacks, websites, and magazines that often traffic in exaggeration, rumormongering, and glamorization of the exploits of grotesquely wealthy drug lords.

To be sure, there are hundreds of clear-eyed writers doing excellent and brave frontline reporting on narcotrafficking and government corruption, maintaining the balance of dispassion while cultivating direct access to primary sources. Gabriel García Márquez's *Noticia de un Secuestro*, a brilliant account of Pablo

Escobar's early-1990s reign of terror in Colombia, was an inspiration: for me it remains the exemplar of how a nonfiction author of the first order—through in-depth interviews, meticulous research, and novelistic technique—can capture the visceral terror wrought by criminals such as the Medellín Cartel.

I was fortunate in this book to have worked with a former federal agent who *lived* it, witnessed it, experienced it all firsthand. It's rare that someone of Drew's caliber leaves a federal law enforcement career at such a young age, while the story of his investigative journey is still so fresh and newsworthy. Together we've strived to write this book with an exacting eye, separating out all the hearsay, rumor, and dubious reporting that surrounded "the world's most wanted narcotrafficker" from the verifiable facts.

All too often stories of men like Drew remain untold. This historic capture operation, with all its remarkable twists and turns, deserves an accurate rendering for posterity. And the key participants—not just Drew, but the other DEA and HSI agents, US marshals, SEMAR troops and commanders—deserve to shine for the years of selfless sacrifice that would otherwise have remained cloaked in shadows.

My deepest gratitude goes to Drew, and to everyone who put in so much hard work—at 3Arts Entertainment, HarperCollins, and ICM Partners—for helping us bring his singular story to fruition.

—D.C.

GLOSSARY

abra las cartas: Literally "open the letters" or "open the books." In the context of a cross-national narcotics investigation, it means "to share all the intelligence."

ARD: Assistant regional director. DEA GS-15 rank in a foreign post.

ASAC (pronounced "EH-sak"): Assistant special agent in charge. DEA GS-15 rank in the United States.

ATF: Abbreviation for United States Bureau of Alcohol, Tobacco, and Firearms. (Now officially known as the Bureau of Alcohol, Tobacco, Firearms, and Explosives.)

AUSA: Assistant United States Attorney.

BDU: Battle-dress uniform. Camouflage fatigues worn by SEMAR.

Beltrán-Leyvas: A Mexican drug cartel run led by five Beltrán-Leyva brothers now based in the northern Mexican state of Sinaloa. Founded as a branch of the Sinaloa Cartel, the Beltrán-Leyvas became their own cartel after the arrest of Alfredo Beltrán-Leyva, a.k.a. "El Mochomo," in 2008, blaming Chapo Guzmán for the arrest.

birria: A spicy Mexican stew dish traditionally made from goat meat.

Caballeros Templarios, Los: A Mexican drug cartel known in English as the Knights Templar, composed of the remnants of the defunct La Familia Michoacana drug cartel based in the Mexican state of Michoacán.

cajeta: Literally "caramel"—DTO slang for high-grade marijuana.

carnal: Often abbreviated in text messages as "cnl," literally meaning "related by blood," it is a term of affection similar to "brother" or "bro."

chanclas: Sandals.

chilango: Mexican slang for residents of Mexico City or people native to the capital city.

cholo: Originally meaning a mestizo, or a Latin American with Indian blood, "Cholo" can now denote a lower-class Mexican, especially in an urban area; a gangster; or in the cartel underworld a particularly tough individual (such as "Cholo Iván" Gastélum, the plaza boss for the coastal city of Los Mochis).

CI: Confidential informant.

Cisne Negro: Spanish for "black swan." The name of the top-secret SEMAR operation to recapture Chapo Guzmán in January 2016.

confidential source: DEA's term for a confidential informant.

cuete (pronounced kweh-TAY): Literally "rocket" in Spanish, is common slang for a pistol or other handgun.

deconfliction: A common law enforcement check to reduce the risk of targeting the same criminals causing a potential "blue on blue" (law enforcement targeting law enforcement) incident.

desmadre: Literally, from "your mother," roughly translated as "mess-up" or "chaos."

DTO: Drug-trafficking organization.

el generente: The manager and a codename for Chapo Guzmán.

El Señor: A term of respect meaning "sir" or "the man," and a code name for Chapo Guzmán.

Gárgola: Spanish for "Gargoyle." The name of the top secret SEMAR operation to capture Chapo Guzmán in February 2014.

G-ride: Short for "government ride." Used by federal agents to refer to their official government vehicle, or OGV.

GS: Group supervisor, DEA GS-14 rank in the United States and in foreign posts.

Guadalajara Cartel: See page 16.

güey (pronounced "whey"): The equivalent of "dude."

halcón (los halcones): Literally "hawks" in Spanish, they are lookouts and street cartel associates who report activities, warning the drug cartels about movements from other DTOs, the police, or the military.

Inge: Short for *ingeniero*, literally "engineer" in Spanish and a code name for Chapo Guzmán.

jefe de jefes: Literally Spanish for "boss of bosses." The name applied to the highest leader of a drug cartel in Mexico and is most frequently associated with Miguel Ángel Félix Gallardo.

JGL: Initials for Joaquín Guzmán Loera.

La Paz: Literally "the peace," a Mexican city located on the southeastern edge of Baja California peninsula.

Las Tres Letras: Literally Spanish for "the three letters." Drug cartel slang for the DEA.

Lic: Short for *licenciado* (see below).

licenciado: Literally "one with a license," it may refer to anyone with a higher degree of education, such as a lawyer, engineer, architect, accountant; within cartel slang, this almost always refers to a lawyer or an educated adviser.

machaca con huevo: A Mexican dish of shredded dry beef that is scrambled with eggs often eaten for breakfast.

mariscos: Seafood, especially shellfish such as clams, oysters, and shrimp, very popular in Sinaloa.

más tranquilo: More calm.

Miapa: Slang for "my dad," and one of the code names for Chapo Guzmán.

mirror: A technique used by drug-traffickers to evade electronic surveillance by law enforcement—most commonly by having texts or messages manually retyped by a low-level cartel employee from one BlackBerry or mobile phone into another (creating a "mirror"), making it difficult for law enforcement to track the messages to the final recipient, and hindering wiretapping efforts.

Nana: "Grandmother"; another code name for Chapo Guzmán.

narco: General term for drug trafficker.

narcocorrido (pronounced "NARkoko'RIðo"): Literally a "drug ballad." An enormously popular subgenre of the Mexican norteño, traditional folk music from northern Mexico. Modern narcocorridos are considered to have started in 1974 with the hit "Contrabando y Traición" ("Smuggling and Betrayal")—the first hugely successful narcocorrido—by Los Tigres del Norte. Today's narcocorrido scene is immensely popular in both Mexico and the United States, with artists taking commissions from real-life cartel bosses and traffickers to celebrate their exploits. With a rollicking beat driven by tubas and accordions—and lyrics often celebrating murder, revenge, and violence—the contemporary narcocorrido scene is often likened to 1990s gangsta rap. It is now arguably Mexico's most popular form of music among young people—often despite a lack of radio airplay and attempts by the authorities to ban the music. The scene is thriving, with artists such as Roberto Tapia, Gerardo Otiz, Movimiento Alteradand, and El Komander drawing huge fan-bases with songs that often celebrate drug lords such as Chapo Guzmán and other high-level traffickers.

narco juniors: The children of the older drug traffickers—a new and often flashier generation of narcos. Unlike their fathers or grandfathers, narco juniors have for the most part been raised in urban wealth, with a higher level of education.

Navolato: A Mexican city just to the west of Culiacán in Sinaloa State.

NCAR: DEA's North and Central Americas Region, covering Mexico, Central America, and Canada.

Padrino: "Godfather," and one of the code names used for Chapo Guzmán.

palapa: A traditional Mexican shelter/structure roofed with palm leaves or branches, especially one on a beach or near a body of water.

pan dulce: Mexican sweet bread often eaten during breakfast.

PF: Policía Federal—Mexican Federal Police.

PGR: Abbreviation for La Procuraduría General de la República, the Mexican equivalent of the Office of the Attorney General, similar to the United States Department of Justice.

pinche cabrón (vulgar): Mexican slang for "motherfucker" or "asshole," it may also be used as a compliment in the sense of someone who is a "fucking badass."

plaza: Territory, turf, or primary smuggling route from Mexico into the United States. May also mean the taxes one must pay to use such routes.

pocket trash: Law enforcement term for anything found left over in pockets—could be receipts, miscellaneous notes, ticket stubs, SIM card remnants, gum wrappers, or anything else.

rápida: Literally "fast" or "high speed," Mexican slang for the armed pickup trucks of SEMAR.

raspados: From the word "scrapes," a cup of shaved ice and sweetened with various fruit juices.

Regional Director: The DEA's highest-ranking senior executive in a foreign post. The regional director is in charge of a foreign region (e.g., the Mexico City Country Office, which covers DEA offices in Canada, Mexico, and Central America). Reports directly to the DEA's chief of operations in Washington, DC.

SAC (pronounced "sack"): Special agent in charge, the DEA senior executive with the highest rank in charge of a specific division office in the United States (e.g., the Chicago Field Division, which covers Illinois, Indiana, Wisconsin, and Minnesota).

Secre: Short for *secretario*, or secretary, and a code name for Chapo Guzmán or his secretaries Condor and Chaneke.

SEDENA (pronounced "sey-DAY-nah"): La Secretaría de la Defensa Nacional, Mexican Army.

SEMAR (pronounced "sey-MAR"): La Secretaría de Marina. Mexican Marines.

sicario: Literally a "hitman" or assassin for the cartels.

Sierra Madre: The major mountain range that runs northwest to southeast through northwestern and western Mexico along the Gulf of California and primarily through the eastern portion of Sinaloa.

straw purchaser: Someone with a clean background who agrees to acquire goods or services—usually illicit—for someone who is unable or unwilling to personally purchase them. These goods or services are then transferred to that person after they are purchased. They are often hired by DTOs and gunrunners.

sugar skulls: Candied sugar in the shape of a human skull, decorated with colorful icing and glittery adornments representing a departed soul or particular spirit on the Mexican holiday known as Día de los Muertos, or the "Day of the Dead."

tacos de canasta: Homemade tacos served out of a basket, frequently from the trunk of someone's car.

tlacoyos: Oval-shaped tortilla pockets made of masa, stuffed with refried beans, cheese, or fava beans, and topped with queso fresco, nopales, and salsa. These are typically served by street vendors off a *comal*.

tolls: Call detail records from a phone.

Zetas, Los: A Mexican drug cartel formed when commandos of the Mexican Army deserted their ranks and began working as the enforcement arm of the Gulf Cartel. In 2010, Los Zetas broke away and formed their own cartel. Considered the most violent of today's cartels, they are also heavily involved in racketeering, kidnapping, and extortion.

INDEX

Page numbers in *italics* refer to maps.

INDEX

ABOUT THE AUTHORS

ANDREW HOGAN is the DEA special agent who led the investigation and capture of El Chapo Guzmán. He now works in the private sector and lives in an undisclosed location.

DOUGLAS CENTURY is the author and coauthor of such bestsellers as *Under and Alone, Barney Ross, Brotherhood of Warriors*, and *Takedown: The Fall of the Last Mafia Empire*, a finalist for the 2003 Edgar Award in the category of Best Nonfiction Crime.

8/2,   6/2/8